P9-BIN-375

Ex Libris
Universitatis
Albertensis

University of Alberta

QUAECUMQUE VERA

THE CORPUS AND THE CORTEX

The Corpus
and the Cortex

JACQUES M. CHEVALIER

The 3-D Mind
Volume Two

McGill-Queen's University Press
Montreal & Kingston · London · Ithaca

© McGill-Queen's University Press 2002
ISBN 0-7735-2357-X

Legal deposit third quarter 2002
Bibliothèque nationale du Québec

Printed in Canada on acid-free paper that is 100%
ancient forest free (100% post-consumer recycled)
and processed chlorine free.

This book has been published with the help of a grant
from the Humanities and Social Sciences Federation
of Canada, using funds provided by the Social Sciences
and Humanities Research Council of Canada.

McGill-Queen's University Press acknowledges the
support of the Canada Council for the Arts for our
publishing program. We also acknowledge the financial
support of the Government of Canada through the Book
Publishing Industry Development Program (BPIDP).

**National Library of Canada Cataloguing
in Publication Data**

Chevalier, Jacques M., 1949–
The 3-D mind
Includes bibliographical references and index.
Contents: 1. Half brain fables and figs
in paradise – 2. The corpus and the cortex –
3. Scorpions and the anatomy of time.
ISBN 0-7735-2355-3 (v. 1)
ISBN 0-7735-2357-X (v. 2)
ISBN 0-7735-2359-6 (v. 3)
1. Neuropsychology. 2. Semiotics – Psychological aspects.
3. Semiotics – Philosophy. 4. Psycholinguistics.
5. Language and languages – Philosophy.
6. Neurophysiology. 1. Title.
QP360.5.C43 2002   302.2   C2002-900769-0

Typeset in Sabon 10.5/13
by Caractéra inc., Quebec City

AUGUSTANA LIBRARY
UNIVERSITY OF ALBERTA

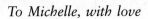

To Michelle, with love

Contents

Contents

THE CORPUS AND THE CORTEX

Log On:
Options and Preferences

The 3-D Mind 1 explored the brain lateralization phenomenon, or the differences observed between right and left brain activity. It also examined effects of bihemispheric integration, or the extent to which the two brains supplement one another, feeding bilateral information into all productions of the "mind." Symbolic materials illustrating these principles were taken from the domain of proper naming practices and from the field of ethnobotany. The latter included plant imagery found in the scriptures (the fig-apron motif in Genesis), in poetry (pines and hemlocks in line 1 of Longfellow's *Evangeline*) and in native Mexican mythology (corn god stories as told by the Nahuas and Popolucas of southern Veracruz). Imageries thus selected for my introductory lessons in "neurosemiotics" were shown to convey cultural views on matters pertaining to history and the human condition.

The central thesis developed throughout these analyses, however, is that sign linkages mapped along the lateral, sagittal, or commissural plane (right and left hemispheres) of the brain or mind add up to constructions of similarities and differences. Semiotic lines of assembly are subject to geometric arrangements of lines of convergence and divergence, assemblages that constitute the divisive and orderly measurements of nervous sign activity. Terms used to capture this thesis revolved around notions of diacritic (mostly left-brain) and syncretic (mostly right-brain) processing. The terms in question are useful by virtue of being "flexible simples": they presuppose inherent interactivity; they do not represent fully fledged tasks (e.g., auditive or visual); and they do not carry legacies of wholesale western philosophy (as in the analytic/synthetic divide).

The broader implication of this argument is that signs do not generate meaning by virtue of their ability to "represent" non-signs, attaching sign-references to objects in the world and ideas in the mind. Instead of being in the business of "referencing," signs engage in "conferencing." As with neurons, they communicate something by communicating amongst themselves. Sign actions *confer* meaning essentially through "signaptic" conversations: meanings are thus interwoven through "reticles" of signs and signals – fine lines of communications evolving in language and in neural cells alike.

Given the complexity of neural and sign reticulation activity, *The 3-D Mind 1* (Chevalier 2002a) made a case for a Deleuzian "chaosmos" governing our "subject-matter" – a brain-mind phenomenon simultaneously mental and biological. It demonstrated how forces of divergence engender wide-open variegations and dispersals of signs in language and culture. The book nonetheless looked for some order lurking behind chaos in the jungle. To wit, my analyses progressively converged on a flexible understanding of the order of similarity and difference, a bilateral mapping of lines that tie signs and hemispheres together. My search for patterns in chaos also led me to pursue a syncretism of neuropsychological, semiotic, and philosophical studies of the mind at work, a malleable logic governing the corpus and the cortex alike.

While relatively ambitious, this overall syncretic-diacritic perspective was developed at some cost to other dimensions of our subject matter. Critical of theories of "spatialized cognitivism" as it may have been (questioning models that assign discrete functions to determinate areas of the brain), *3-D Mind 1* paid attention mostly to cognitive activity mapped on to the lateral plane (right hemisphere – left hemisphere). Lines of divergence and convergence may be very different, yet they belong to the same commissural axis. This is the axis where cognitivism rules, albeit the "nervous" kind, the bihemispheric sort that allows "chaotic and decentred" effects to play themselves out. All is as if my reading of half-brain talk and related models acted as a foil to a higher form of cognitivism, a superior logic converging on the integral and differential calculus of *syncretic and diacritic* processing. The spectre of a well-tamed "savage mind" ruled by Saussurian and Lévi-Straussian notions of synchrony and diachrony loomed large in my discussion of neurons and signs.

Thus some important aspects of the nervous sign system were left unattended. Noncognitive operations were deliberately relegated to a secondary role. This raises the following question: aside from its residual connotations, what do we mean by the "noncognitive" dimension of language and braim? What is the "other side" of the logic of convergence and divergence that deviates from cognitivism broadly defined (using input from both hemispheres)? If by *synkretismos* is meant the union of two parties against a third, what third party is excluded from what I coined a "theoreticle" perspective on the two-sided weavings of similarities and differences?

Interestingly, questions about conflicting perspectives on sign and synapse contain their own answer. They point to an issue not adequately covered in *3-D Mind 1*: the status of conflicts and battles in language. Signs in conflict involve lines of logical dissonance or divergence. But disputes in language involve other considerations as well, such as the emotive and normative (moral, rational-instrumental, interpretive) underpinnings of language and brain activity, and the corresponding application of force and intensity designed to enforce proper sign attention. Signs are not merely lines that fulfil divisive and connective functions. They also draw lines of battle between good and evil, right and wrong, the pleasurable and the unpleasurable, the practical and the impractical, the lawful and the lawless. In keeping with these struggles, semiotic lines are prone to separate winning signs that must be voiced from those that lose out and must be silenced. Signs that win and receive a good hearing convey meaning through lines to be read and stuff to be read between the lines. They generate explicit and implicit meanings, actively speaking against signs that are squarely out of line. In the end, some sign actions deserve attention, others less or none.

Attentional polity pursued for emotive and normative ends is at the heart of sign activity. The rule applies to all meaningful assemblages, including the pronouncements of grand theory. When giving attentional preference to one dimension over another, say, the rhizomatic (the tropical jungle) over the arborescent (the gardens of Versailles), as in Deleuzian philosophy, theorists exercise choice and use persuasive and rank ordering techniques entailing attentional and rhetorical devices of all sorts (e.g., elevating the convoluted rhizome above the logical tree). However, this begs the question: Given the inevitable tactics of persuasion, what role and weight should we assign to attentional politics in brain and language activity? How do signs in battle and the uneven allocation of attention affect reticles of convergence and divergence?

Answers to these questions lie outside the cognitive domain proper. They point to the investments of emotive and normative (moral, rational-instrumental, interpretive) within brain and sign activity, the subject matter of this book. Note that the previous book has already introduced the issue of emotional attention. We saw how emotionality connects to the brain lateralization phenomenon. While the RH (right hemisphere) is often characterized as the emotional brain, our reading of the evidence suggests that both hemispheres give shape to human feelings and sentiments. The question is whether there is more to emotion and affect than mere attributes of right and left brain activity. What is emotionality in the first place? Is it a function reducible to effects of cognitive mapping, as in dissonance theory? Is it a physiological experience that may or may not be reflected in consciousness and knowledge? Is it a distinct faculty or function that can be located in one or several areas of the brain? Alternatively, could it be just an arbitrary construct, a term falsely declined in the singular, a fictive construct that has its own history and that occupies no distinguishable terrain in either language or the brain?

The 3-D Mind 1 also raised issues of attentionality proper – the force that brings sign actions to our awareness. Remarks were made regarding right-brain involvement in the semiconscious attentions of dreaming, implicit learning, and unfocused vigilance, which is the readiness to detect and respond to events that are unpredictable or rare (Robbins 1998: 190, 199). These observations are indicative of variations in degrees of attentionality. Thus the RH seems particularly responsive to things we may not need to consider explicitly; it "attends" them without using maximum awareness. Again, these comments raise important questions that need further exploration. How do mechanisms of implicit impressions and explicit expressions function at both the neurological and semiotic levels? Why is it that sign activity does not simply spell everything out? Why does it let some signs accede to the high road of attentionality (or is it "consciousness"?) and channel others through the low road of implicitness or sheer unawareness? How do these uneven attentions feed into productions of feeling and judgment?

This book tackles such questions via studies of emotive and normative attentionality. The corresponding mechanisms are unpacked and shown to constitute what might be called the axial plane of our subject matter. This is the plane of vertical projections that connect the lower and upper structures and functions of language and brain, hence signs of limbic affect (subcortical) and prefrontal judgment (cortical). Some of these vertical projections crept into the semiotic analyses of *3-D Mind 1*. Signs were shown to convey impressions and expressions of things deemed pleasurable or unpleasurable on the plane of desire (the limbic system), and also things viewed as commendable or reproachable on the moral plane (the prefrontal system). But the implications of these vertical projections connecting the two systems have yet to be addressed. Three examples of normative and emotive attentions in language are briefly explored below, imageries taken from *3-D Mind 1* and not explored at greater length in other sections of this book. They include proper naming practices, the hemlocks and forest primeval imagery in Longfellow's *Evangeline*, and the fig apron motif in Genesis 3.

SIGNING ONE'S NAME

Consider my foray into proper naming practices, using my own signature (Jacques M. Chevalier) as an illustration of "sign composition and conferencing" in language. Although unstated, one argument follows directly from my analysis of similarities and differences between "Jacques" (Christian name) and "Chevalier" (biological patronym): a simple sign action such as signing one's name conveys a logic of desire, a package of norms and morality that goes beyond the name's denotative function and the categorical grid sustaining it (e.g., proper nouns and common nouns forming two separate categories). In the case of this author's signature, rules of proper naming make it clear that one's Christian name comes and ranks first and should be elevated above the "altered" surname (French *sur*, over, and *nom*, name). "Jacques" evokes the life of an apostolic spirit, to be valued above the ephemeral attributions and affiliations of biology and the family patronym. But with the surname and recognized family tie come also a whole set of legal claims and entitlements instituted through family law. Implications of intellectual property and accountability attached to an author's name are another site of moral expectations embedded in acts of signature. These normative underpinnings of proper naming practices suggest that there is more to signing one's name than simply designating one's identity, using a conventional practice to mark out a person merely for purposes of address or reference.

A strictly cognitive analysis of explicit and implicit signs embedded in "Jacques Chevalier" would leave aside the moral order erected through "proper" names. But it also neglects the fact that uneven attention is often granted to different parts of one's name. This brings us to yet another feature of naming: signs of impropriety that are left unnamed according to circumstance. Given its composite nature, a proper name can be broken down into malleable parts that can be emphasized in variable ways and with adaptive import. When celebrating my funeral, the officiating priest called upon to evoke my memory will in all likelihood use my Christian name and refrain from uttering my surname, custom *oblige*. This he will do not with the purpose of refreshing everyone's memory regarding my identity as the person being mourned. The omission will point rather to a will to focus on the life of the spirit, or the soul that transcends the body and survives after death. Since the event takes on a religious character, emphasizing the surname would be improper. Given other circumstances, addressing a living acquaintance by his or her given name alone will also serve a meaningful purpose: expressing the ethics of familiarity or a wish thereof. Conversely, there are situations such as court procedures where addressing or referring to persons by their first names is frowned upon as a sign of inappropriate informality; the first name is better left unsaid (as in "Cher Monsieur Chevalier"). Even when things and people are "simply" called by their names, the order of morality and desire weaves in acts of silence where inattention is called for.

MURMURING PINES AND HEMLOCKS

The primeval forest imagery borrowed from Longfellow can serve to further illustrate the normative and emotive weavings of sign activity. We have seen that the hexametric prosody and the first line of *Evangeline* ("This is the forest primeval, the murmuring pines and the hemlocks") convey a mood of plaintive monotony, in keeping with the Acadian story of forced deportation and paradise lost (Chevalier 1990). The hemlock motif is instrumental in strengthening this mournful affect. It conveys a "spirit of lower gravity" (branches drooping, weed growth) coupled with notes of bitter death (by poisoning) and a "final end" utterance – a two-accented spondee that puts a stop to a line of dactyls consisting of three syllables each, one accented followed by two unaccented.

This analytic reading of the overt meanings and rhythmic structure of sign utterances sheds light on sentiments and feelings evoked in language. However, an interpretation of the formal-analytic kind neglects a crucial property of the sign process: the fact that some signs are given greater attention than others. The exercise leaves aside a key feature of the "murmuring pines and the hemlocks" composition: the fact that signs are heavily biased toward the brighter evergreen forest imagery. The darker indices of the hemlock motif are prevented from entering the overt text; sombre implications have a negative impact on the composition only by way of indirection. Likewise, loftier usages of the hemlock, those embodied in stories and morals of Christ's sacrificial wormwood (Christ swallowing a mixture of wine and gall immediately before his death), are barred from entering the surface narrative. Given the affects they provoke, some meanings are bracketed while others are brought to the fore. The interplay between the overt and the covert – between the explicit, the implicit, and the illicit – is part and parcel of the logic of sign activity. This is a logic loaded with the battle stakes of morality, conceptions of the good life, and the embattlements of cultural history.

Rather than being conveyed through sign meanings alone, affects are generated through tactics of attentional manipulation. These manipulative operations determine the overall economy of words, which includes signs that are uttered but also those that are actively invested with little or no attention. Manipulative techniques are so pervasive that they will also govern the rational-instrumental aspects of sign activity. The prosody of Longfellow's *Evangeline* is a good example of this. Hexametric technology is a rational-instrumental use of techniques and "norms" of language aimed at achieving particular literary ends. But hexametric measures are also shot through with "normative-moral" affects that take position in the battlefields of history.

The English dispute over the use of hexametric prosody (Guest 1968) proves the point. Briefly, one characteristic of the English hexameter is that it must harmonize rhythm and quantity with the relative length or brevity of vowels, especially the accented ones. The problem with English syllables, however, is that they must be made stronger or weaker than they really are in order to meet the classical requirements of the Greek or Latin measure. These technical problems account for the resistance met by the classical hexameter throughout the history of English literature. Such difficulties are not found in German poetry, which is essentially accentual, nor in classical verse, which works quantity and caesura alone. Given these obstacles, British and American poets of the nineteenth century generally held the hexameter to be alien to the English language and "genius." Their preference went instead to the iambic pentameter, in keeping with Samuel Daniel's *A Defence of Rime* first published in 1603. In their eyes hexametric poetry was a subversion of Anglo-Saxon culture masquerading as a contribution to Europe's classical Renaissance (*à la* Coleridge and Southey).

This battle over the hexameter points to broader developments in nineteenth-century literature, ethics, and philosophy. The revival of the hexametric genre implied a measure of reversion to cultural paradigms lost to modern history. *Evangeline* was written at a distance from the national tradition, closer to what some considered an older and more "natural" mode of expression. The poem dealt sympathetically with a "primitive" people of Latin descent subjected to the British rule. Accordingly, the ancient form it employed introduced a Latin measure into an English literature confined to its pentametric tradition. In keeping with the spirit of romanticism, the metre embodied a measure of disenchantment with limitations and formal strictures imposed by modern history.

English writers of hexametric poetry were thus playing for high stakes. They sought to invest words of poetry with norm, mood, and affect designed to do two things: moving individual readers in certain ways (nostalgia for the Golden Age), and pushing national literature in directions at odds with established conventions (closer to nature and its primitive ways). Sign techniques never exist for their own sake. They are deployed primarily by virtue of the attentions and gains they pursue.

FIGS IN GENESIS

This argument regarding the battlefield of semiosis applies equally well to other botanical motifs explored in *3-D Mind 1*, imageries taken from native corn mythology and the fig apron scene of Genesis 3. This book will not dwell on the normative and emotive underpinnings of corn and fig imageries. Nahua corn imagery will be revisited in *3-D Mind 3* (Chevalier 2002b). The ways in which native symbolism raises questions of morality and transgression will be taken up together with issues of temporality in language.

Norms and affects embedded in the fig motif have already been addressed. In the previous book the fig apron motif was viewed as a complex sign field that can trigger one of four impulses, a "sign action potential" that must leave three other connections "at rest." The four connections are of the joyful, the sinful, the tribulational, or the sacrificial kind. The scene of Adam and Eve hiding their sexuality with leaves of the fig tree – a self-sterile tree that bears edible fruit on account of wasps carrying pollen from the male caprifig and dying inside the female fig – pays explicit attention to the sinful and tribulational implications of the fig imagery. It emphasizes evocations of reproductive sexuality and related trials (man separated from woman, life from death) of the human condition. By implication, other possible directions are excluded from pronouncements of the overt script, written off by the "central sign system" as it were. But we shall see in *3-D Mind 3* that hopes invested in alternative usages of the fig motif, be they joyful or sacrificial, are bracketed and held in abeyance at best. The scene of man and woman wearing an apron of fig leaves suspends other teachings of the fig motif, such as man enjoying a good life in the shade of a fig tree, or Zacchaeus renouncing his wealth and climbing into a fig tree for a sight of the saviour. Although not made explicit in Genesis 3, supplementary meanings of the fig point to sign associations at rest, hence other connections that form an integral part of broader biblical stories about what comes after the Fall. More shall be said on the hermeneutics of prefiguration later. For the moment I emphasize the bracketing of signs and related anxieties not explicitly attended by the surface narrative, a central topic explored throughout this book.

This book expands on the emotive and normative dimensions of sign tactics and attentions. It does so against the background of studies in neuropsychology, semiotics, and philosophy. Detailed illustrations of the embattlements of sign activity are provided, together with discussions of the philosophical implications of the embeddedness of morality and desire in semiosis. Material involving animal symbolism (frogs and beavers) and bodily imagery (hips, feet, shoes, body piercing) are examined in the light of the broader issues outlined above. The interpretations I offer converge on the notion that not all things signified are given full attentionality. They also point to some of the basic reasons and mechanisms that account for the axial mapping of sign projections and attentions in language.

A major difficulty encountered in this area of research – "depth studies" of signs of the overt and the covert – lies in cognitive approaches to the unconscious. An axial or vertical perspective on semiotic activity must be careful not to simply substitute notions of deep-seated logic for older imageries of God – the nonperceptible Logos or Spirit that puts life into the body. With cognitivism the invisible Spirit that rules beyond the material sphere becomes logic hiding and ruling from inside the visible corpus: in other words, the unconscious Code. Our discussion of the normative and emotive *attentions and inattentions* of nervous sign activity departs radically from these structural views of *l'esprit* (the mind, in French).

Cognitive models of the "unconscious mind" are reminiscent of Descartes's grammatical analysis of the sentence "Dieu invisible a créé le monde visible" ("God the invisible created the visible world"), as discussed by Chomsky. Chomskian linguistics is an important contribution to a philosophy of language stripped of theological influence. In his theory of deep-mental operations that generate surface-linguistic structures, Chomsky nonetheless reproduces a basic principle of Cartesian philosophy: the invisible plays a "creative" role. The invisible is no longer the Creator ruling over his Creation. It becomes rather the implicit-mental logic of propositions (e.g., "*Dieu est invisible*" + "*le monde est visible*" + "*Dieu a créé le monde*") that govern the explicit-verbal judgment of language (e.g., "*Dieu a créé le monde visible*"). From an ethereal Spirit the Sign-Maker is turned into an incorporeal mind, a coding machine that presides over all sign-manifestations of *l'esprit*.

The question raised in this book is whether a dialogue across disciplines can help us do away with the legacies of Logos. Can we speak of "deep structures" that are invisible in the sense of not being expressed, without falling back on some structural-linguistic version of the unconscious "spirit"? Can depth-analysis escape the notion that unspoken semantics are abstractions "represented in the mind"? Can we avoid speaking of thought or the unconscious as a non-thing that acts invisibly, like a "mental accompaniment to the utterance" (Chomsky 1966: 34)? Should we allow connections generated "deep in the mind" to be expelled from the physical brain, as if thoughts could be disembodied and take "leave of their senses"? Can we not argue instead that the invisible is as material as everything else? Should we not recognize "unaware semiosis" for what it is: a particular set of methods for "doing things" in language and the brain?

Some answers to these questions may be found in what we might call a *homunculus* (the Latin diminutive of *homo*) conception of semiosis. The alchemists once claimed they could create a "little man" endowed with supernatural power. In keeping with western views on spirits (and consciousness for that matter), their homunculus was designed to have no body or sex. Modern medicine has performed radical surgery on this dwarf. It has stripped him of his supernatural powers and restored a physical form to the creature. The homunculus nonetheless managed to keep some monstrous features. He is now used as a model that captures central characteristics of human biology, features that are often unexpected and do not reflect common perceptions of the body. The motor and primary somatosensory homunculus exemplifies this (Rosenzweig et al. 1999: 207, 301). The man is still small, for reasons of convenience. But he also has huge hands and lips. Actually the size

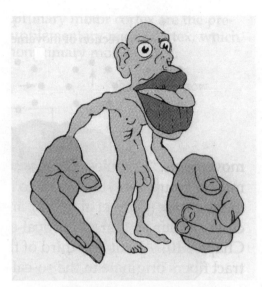

The homunculus. Rosenzweig, Leiman, and
Breedlove, *Biological Psychology*, 301

of his body parts is radically out of scale. Instead parts are
proportional to the amount of cortical areas serving them. The
model is now the embodiment of an uneven attention principle
mapped on to the body and waxed into the brain.

The task I undertake in this book is to perform a similar and
closely related operation on consciousness – giving it body parts
(private ones included) with variable proportions that correspond to the uneven attentions it receives from nervous sign
activity. The end product is a "little man" endowed with an
attentional algorithm scattered over the "larger intelligence" of
his body, using procedural forms that are not under his executive control (LaBerge 1995: 214). To understand this algorithm, we first turn to neuropsychological studies of emotive
and normative (moral, instrumental, interpretive) aspects of
brain activity.

NEURAL MINDFULNESS

Synaptic Attentions
and Reticular Activity

Neuropsychology no longer permits us to think of cerebral hemispheres as cognitive modules unaffected by human feelings and the unconscious. Half-brain studies suggest that each hemisphere is emotionally profiled, especially the right one, which when impaired leads to deficits in nonverbal emotional expression. More importantly, studies of cortical and subcortical processes show how hemispheres are affected by emotions involving brain circuits and systems that proceed along lines other than the strictly hemispheric. The same can be said of attentionality and unawareness. They have impact on hemispheric differences, yet they too result from neural mechanisms and systems that go beyond the brain lateralization phenomenon.

What are the neurobiological foundations of affect and awareness? How do emotions interact with attentionality? We know that when they come together, emotion and attentionality produce awareness of feeling. The problem with this observation, however, is that the body and brain can register information and emotions such as fear and anxiety without the subject's full cognizance. Not all effects of brain activity accede to what we call "consciousness." This raises an intriguing question: how can neurons convey information that never comes to our attention? How can neural circuits learn anything without letting the mind become aware of what it is learning?

In an attempt to answer these questions, I review four ways in which nervous communications can operate in the absence of awareness. All four ways speak to both the distance and the ties that lie between cognition and emotion, or "consciousness" and the body's "larger intelligence."

First, some neural *depolarizations* permitting the propagation of nerve influx are achieved without activating the neocortex. The brain may achieve awareness of some of these depolarizations. As explained below, effects of cerebral arousal and vigilance nonetheless presuppose a complex summation of "synaptic inattentions" – *polarizations and hyperpolarizations* that maintain or increase levels of membrane resistance so as to bar alternative pathways from transmitting competing signals.

Second, the *reticular system* plays a crucial role in arousing the brain from sleep or inattention. This system resembles sensory processing (Rosenzweig et al. 1999: 191) in that it works selectively and restrictively, screening incoming stimuli for relevance and immediate sensorial and behavioural attention.

The other two mechanisms will be considered later. One has to do with *limbic emotions* and feelings (reticle 2). These affects may feed into "higher mental" functions such as prefrontal judgment and goal-oriented behaviour. But they may also be kept in check by the prefrontal brain or be processed independently of cortical attentionality. When the latter occurs, subcortical activity generates *implicit learning, feelings, and emotive events* channelled through inter- and intrahemispheric pathways. The last source of "active unawareness" stems from the fact that "consciousness" has no direct control over neural exchanges of the peripheral autonomic system. This is the system that either stimulates muscular and skeletal movement through sympathetic pathways or inhibits it through parasympathetic action (reticle 3).

What we call "consciousness" is the tip of a neural iceberg that hides more than what it actually reveals. Inhibition, inattention, and resistance are basic habits of neural mindfulness.

SYNAPTIC ATTENTIONS

Consider first the mechanisms of communicational attention and inhibition occurring at the synaptic level. Neurons consist

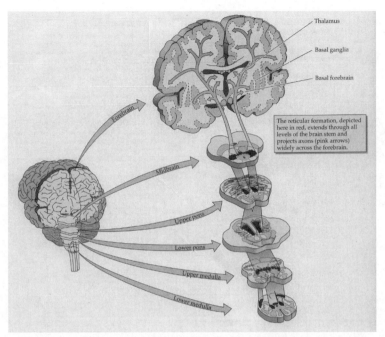

The reticular formation. This system is thought to activate the rest of the brain. Rosenzweig, Leiman, and Breedlove, *Biological Psychology,* 396

of cell bodies that when triggered transmit an electrical charge down the cordlike axon. The charge stimulates the vesicles of chemical neurotransmitters stored at the axon terminals. Variable cocktails of neurotransmitters are then released and come into contact with the dendrite tentacles extending from the cell body of another neuron. The end result is a synaptic stimulation or nervous impulse that comes from one neuron and that either "fires off" or inhibits another neuron. Alternatively, direct synaptic contact may occur between the dendrites of different nerve cells (some of which are without axons); between their axon terminals; or between presynaptic dendrites and the axon hillock area located at the top of the postsynaptic axon.

Impulses are a combination of chemical and electrical cell events. When the negative and positive poles of a battery are connected and the circuit is closed, energy stored in the battery

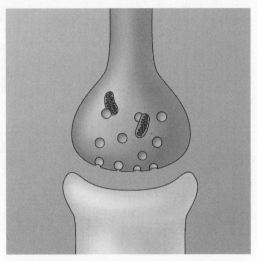

Synaptic transmission. Rosenzweig, Leiman,
and Breedlove, *Biological Psychology*, 32

is released. Electrons flow from the negatively charged area to
the positively charged area. Oppositely charged areas that are
kept separate have a potential energy measured in volts or millivolts. The same reasoning applies to neurons. The membrane
that separates inside and outside cell areas maintains a difference of potential and thus stores electrical energy; this property
of the membrane is called *capacitance* and can be quantified.
The more pronounced the difference is between areas that have
opposite charges, the greater is this difference in membrane
potential. The normal potential registered between inside (negative) and outside (positive) areas of the membrane is about
−70 millivolts (mV). When the two areas are kept separate, the
potential is said to be at rest (Vm) and the membrane is polarized, with a voltage that varies from -40 to -90 mV. The liquid
that lies outside the membrane contains most sodium (Na+) and
chloride (Cl -) ions. By contrast, the intracellular cytoplasm
liquid contains greater concentrations of potassium (K+) and
negatively charged protein ions. The negative resting potential
of the membrane is principally caused by the large, negatively

charged protein molecules trapped inside the cell membrane, and the tendency for positively charged potassium ions to move out of the cell towards regions of lower concentration. Positive ions stop flowing out from the cell when their movement towards areas of lower concentration is cancelled out by their attraction to intracellular negative ions.

The human organism responds to stimuli through nerve impulse, which is a force acting suddenly on a body, or the interval of time during which it acts, and the resulting excitation generating physiological activity. We have seen how nervous impulses involve communications between the terminal bulbs of presynaptic neurons and the cells and axons of postsynaptic neurons. The response at any given site of the postsynaptic cell membrane may result in a depolarization of the membrane potential. This is brought about by changes in the permeability of the membrane permitting a brief influx of sodium ($Na+$), reducing the negative charge inside the membrane (e.g., from -70 to -50 mV). When a membrane area is depolarized, positive ions inside the excitable membrane (mostly $K+$ potassium) will immediately migrate towards negative zones via the cytoplasmic liquid acting as a conductor. Likewise, negative ions expelled from inside migrate towards positive zones. This chain reaction known as the graded (or local) potential depolarizes neighbouring areas, passively spreading the current along the membrane.

A graded potential may generate local activity only, as opposed to firing off a cell and producing a nervous influx. Since the membrane is porous, part of the charge is lost along the way and the potential diminishes as it moves away from its point of origin. Unlike the digital, all-or-none action potential, the graded potential is an analog signal generating continuously varying values. Although short-distanced, the graded potential may turn into a nervous influx relayed by the axon and terminal bulbs, provided that the sensorial or chemical stimuli and the graded potential are "liminal"; that is, a nervous impulse is transmitted when the current is strong enough to reach the hillock along the membrane (the area where the axon emanates from the soma). When a critical threshold is reached, the hillock depolarizes by a certain amount known as the firing level (from −70 mV down to −55 and −50mV).

The nervous impulse, also known as an *action potential* or excitatory postsynaptic potential (EPSP), modifies the cell membrane permeability by opening voltage-dependent channels on the axon membrane. More sodium ions come inside the cell, which becomes even less negative and depolarizes the axon membrane even further, causing lateral displacements of depolarized ions towards polarized areas. Like a domino effect, the nervous influx follows its electrical course along the axon and is passed on to other neurons, without requiring further chemical action or stimulation.

As more sodium ions come inside the cell, the membrane potential may go up to +30 mV during the space of a millisecond. Sodium channels are then closed and potassium ions are expelled from the cell, which becomes negative again. The cell's resting potential is restored and the membrane repolarized through the normal activity of thousands of sodium and potassium pumps.

Cell events are variable. We have seen how excitable membranes can be polarized, depolarized, or repolarized. The type of event depends on what receptor sites or channels are being tapped and also which neurotransmitter is being released. Some neurotransmitters convey inhibitory messages: for instance, GABA (gamma-aminobutyric), which is estimated to be active in no less than 40 per cent of all brain synapses. Membranes, however, can also be hyperpolarized, an event that entails an increase in membrane potential.

This brings us to inhibitory events and functions of neural activity. Unlike excitatory neurotransmitters such as dopamine, inhibitory neurotransmitters such as serotonin produce a hyperpolarization of the postsynaptic membrane by increasing its intake of chloride and outflow of potassium. This increases the relative negative charge inside the membrane (e.g., from −70 to −90 mV). Since hyperpolarizing potentials diminish the likelihood that a nervous influx will be generated, they are called inhibitory postsynaptic potentials (IPSOs). Synaptic events of this type are particularly important in that they serve to reduce noise and focus whatever signal is being transmitted through an action potential. They can also serve to enhance contrasts between central and peripheral stimuli, that is, between highly stimulated neurons and their immediate neighbours. At the same time as sensory neurons excite other neurons through action potentials, they inhibit neighbouring sensory neurons, decreasing their firing rate by some specific amount. Effects of lateral inhibition are commonly found in the activities of nervous tissues, the retina, and the olfactory and gustatory epithelia (Churchland 1986: 54, 73, 76, 135–6).

Thousands of depolarizing and hyperpolarizing exchanges can occur simultaneously in any given synaptic zone, and it is their mutual interaction that will add up to either an excitation or an inhibition of the postsynaptic neuron. "Potentials therefore interact as currents sum to create a large depolarization; or, if the effects were hyperpolarizing, to prevent depolarization; or, if the effects are opposite, to interfere and cancel" (ibid.: 54). The amplitude of synaptic potentials depends on stimulus size and the overall summation of depolarizing and hyperpolarizing potentials, hence overall changes in negative and positive charges recorded across the membrane. The summation implies a subtraction between excitatory and inhibitory postsynaptic potentials and computes both events occurring across the cell body (spatial summation) and over the space of a few milliseconds (temporal summation). To assess the intensity of the nerve impulse, the brain will also rely on its frequency, the quantity of chemicals released by presynaptic neurons, and the array of receptors and nerve cells involving a wide distribution of firing thresholds (Rosenzweig et al. 1999: 195).

When released, a membrane potential and corresponding voltage can be dampened by resistance factors that are either strong or weak. The cell membrane is a powerful insulator, whereas the fluids inside and outside the membrane are conductors characterized by low resistance properties. A current (I) is proportional to the voltage (V) recorded across two separate areas and is inversely proportional to resistance factors (R) at work. The Ohm law thus states that current equals voltage divided by resistance: $I = V/R$. An immediate corollary of this is that when the current is high despite strong resistance factors, the potential measured in volts is proportionately higher (or $V = I \times R$).

When polarized (at rest) or hyperpolarized (inhibited), inter-neural potentials are in a state of *synaptic inattention*, so to speak. External or internal stimuli may still cause presynaptic axon terminals to release chemical signals, but their frequency or intensity is not sufficient to generate communicational impulses and the neocortical attention that may accompany them. Note that a synaptic zone "at rest" is not inactive. A resting potential will require the constant chemical activity described above in order to maintain itself in the postsynaptic zone. "Thus, maintaining the membrane potential, which is nec-essary for the neuron to be ready to conduct impulses, demands metabolic work by the cell. In fact, most of the energy expended by the brain – whether waking or sleeping – is thought to be used to maintain the ionic gradients across neuronal membranes" (ibid.: 1999: 58).

RETICULAR ACTIVATING SYSTEM

Synaptic polarizations, depolarizations, and hyperpolarizations point to micro-level attentional mechanisms triggered or inhib-ited by the nervous system. Depolarizations, however, are not sufficient to generate full cortical attentionality. Essential to complex brain activity as it may be, a nervous influx does not always translate into cerebral awareness. A multitude of neurons can fire without the cortex paying much attention to these events. What is commonly referred to as "consciousness" is nonetheless only a particular kind of neural activity superim-posed on other nervous processes. Far from being a unitary system (Robbins 1998: 190; LaBerge 1995: 216–17), "con-sciousness" refers to several forms of cerebral activity. It involves extensive areas of the cortex and consists in variable levels of attention and arousal, ranging from vigilance at one end of the spectrum to coma at the other end.

The reticular system plays a particularly important role in drawing brain attention to neural activity. This is the brain-stem unit that sorts and records sensory input from the body and controls arousal and cycles of sleeping, dreaming, and waking (Peters 1995: 203–4). It is prominent in the midbrain and extends from the thalamus to the brain stem medulla. It is thought to activate the basal ganglia, the thalamus, and the basal forebrain, which in turn release the dopamine required for brain arousal. Serotonin has the opposite effect, which is to turn the reticular system off and put the brain to sleep (note that lack of serotonin correlates with aggressive behaviour). Injuries to this system can cause severe sensory neglect and persistent sleep (Heilman 1995: 223–4). When active, however, the system works selectively and deploys its own mechanisms of inhibition and inattention. The reticular system is thus responsible for preventing sensorial information from reaching "consciousness," putting aside at least 99 per cent of what our senses actually perceive. This is the only way the brain can prevent too much sensory information from overloading normal cognitive processes.

Studies of reticular system input into brain attentionality are all the more important as there is ample evidence to suggest that early brain development occurs along the axial dimension, not along lines of hemispheric specialization. Using Lurian theory, Satz and others argue that vertical specialization evolves as the child matures, with the implication that some learning disabilities can be linked to problems of subcortical-cortical communications. Development would begin with the primary cortical area and the reticular system. Progression would then proceed to the secondary area responsible for organizing and coding sensory information, and the tertiary area, where information from multiple sources is integrated and fed into the process of planning and executing complex behavioural responses. Boliek and Obrzut add that abnormal patterns of subcortical-cortical development and related activation and arousal deficits may be characteristic of learning-disabled subjects, resulting in problems of lateral specialization and bihemispheric communications. Thus "for a large subset of learning-disabled children, atypical cerebral organization and unusual patterns of laterality, attention, and arousal may underlie deficits in auditory and visual, language, and nonlanguage information-processing abilities" (Boliek and Obrzut 1995: 642; see also Bruder 1995: 665).

The story of brain attentionality starts with synaptic events and reticular activity. But it does not stop there. Ascending reticular system axons project widely from brain stem into forebrain, midbrain, and the limbic system. As we are about to see, these regions also make critical contributions to attentional activity.

Frontal Lobes, Limbic Processing, Implicit Learning

THE LIMBIC SYSTEM
AND THE FRONTAL LOBES

Before I say more about the workings of attentionality, a few words should be said about basic divisions of brain anatomy. Briefly, the human brain is divided into the hindbrain (rhombencephalon, includes the brain stem), the forebrain (prosencephalon), and the midbrain (mesencephalon).

The *hindbrain* includes the pons, the cerebellum, the two cerebellar hemispheres, the medulla oblongata, and the fourth ventricle.

The *midbrain* connects the forebrain and the hindbrain and is composed of (1) the reticular formation; (2) two superior and two inferior colliculi involved in the visual and auditory systems, respectively; and (3) the mamillary bodies, implicated in memory (together with the fornix and the hippocampus).

The *forebrain* represents about one-third of the cerebral cortical surface and consists of two convoluted cerebral hemispheres, which contain the basal nuclei and the amygdala. Beneath these hemispheres lies the diencephalon which includes (1) the optic chiasm; (2) the third ventricle filled with cerebrospinal fluid; (3) the thalamus acting as a relay station between the optic chiasm and the visual areas of the cortex and also between the information going in and out of the cerebral hemispheres; and (4) the hypothalamus (see below).

The *limbic system* crosscuts these divisions. It comprises the anterior nucleus of the thalamus, the hypothalamus, the hippocampus, and the amygdala. To these should be added interconnecting fibre tracts such as the fornix and the mammillothalamic tract.

The *thalamus* is an egg-shaped, major integrating centre where all sensory tracts converge (save for the olfactory); sensory signals are processed here and then passed on to the relevant cortical area. Thalamocortical pathways are found in virtually all areas of the cerebral cortex and play a key role in the management of attention (Laberge 1995: 222) .

The *hypothalamus* is involved in regulating basic homeostatic activities such as eating, drinking, and regulating body temperature. It acts on endocrine glands, connecting bodily reflexes and cortical activities that serve to express motivations of hunger and thirst. The hypothalamus also modulates a person's mood. It is directly at work when pain, pleasure, pulsions (hunger, thirst, sexual drive), and emotions are experienced and expressed physically. If fear is experienced, the hypothalamus will prime the sympathetic system into fight or flight action, producing effects ranging from palpitations to high blood pressure, paleness, sweating, and dryness of mouth. All in all, the hypothalamus monitors autonomic and emotive functions simultaneously, with the implication that emotional stress can result in psychosomatic dysfunctions.

The *hippocampus* is responsible for associating contextual and long-term narrative memories with like-dislike signals coming from the amygdala.

An almond-shaped *amygdala* is located on each side of the brain, immediately above the brain stem. It fulfils emotive functions, causing emotional blindness (inability to feel and recognize feelings) if impaired or severed from the rest of the brain. "Animals that have their amygdala removed or severed lack fear and rage, lose the urge to compete or cooperate, and no longer have any sense of their place in their kind's social order; emotion is blunted or absent. Tears, an emotional signal unique to humans, are triggered by the amygdala and a nearby structure, the cingulate gyrus; being held, stroked, or otherwise comforted soothes these same brain regions, stopping the sobs. Without an amygdala, there are no tears of sorrow to soothe" (Goleman 1995: 15).

Most of the information coming through the retina or the ear goes first to the thalamus. It then goes to relevant neocortical brain circuits for "conscious" analysis, cognitive assessment, and appropriate response. If the stimuli are thought to have emotional content, the neocortex will signal this to appropriate areas of the amygdala (e.g., auditory and visual information is projected to the lateral nucleus). All incoming sensations and perceptions are thus filtered or inspected by the amygdala for emotional valence.

But not all signals triggering the amygdala come from the neocortex. Emotional appraisal by the amygdala can be solicited without cortical input and awareness. If so, where do the signals come from? The answer lies in relatively autonomous functions of the limbic system. Studies by LeDoux have shown that there is a small bundle of neurons that goes directly from the thalamus to the amygdala for immediate processing and emergency reaction, shortcutting the main route up the neocortex. Thus "in the first few milliseconds of our perceiving something we not only unconsciously comprehend what it is, but decide whether we like it or not; the 'cognitive unconscious' presents our awareness with not just the identity of what we see, but an opinion about it" (Goleman 1995: 20). The inspection is done very quickly, in about half the time that the brain would otherwise take to transfer information from the thalamus to the amygdala via the longer neocortex route. This direct precognitive evaluation is done without precision and relevant contextual analysis based on memories of similar events. Although subject to error and prone to triggering anxiety feelings and impulsive behaviour, this fast-track appraisal of emergency signals is essential for the body to respond to emergency situations that require immediate reactions of flight or fight.

Under stress (or anxiety, or presumably even the intense excitement of joy) a nerve running from the brain to the adrenal glands atop the

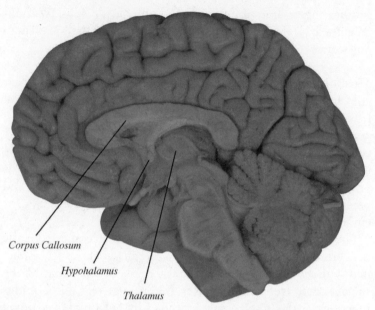

Corpus Callosum

Hypohalamus

Thalamus

The thalamus and the hypothalamus. Sundsten, *The Digital Anatomist*

kidneys triggers a secretion of the hormones epinephrine and norepinephrine, which surge through the body priming it for an emergency.
These hormones activate receptors on the vagus nerve; while the vagus
nerve carries messages from the brain to regulate the heart, it also
carries signals back into the brain, triggered by epinephrine and norepinephrine. The amygdala is the main site in the brain where these
signals go; they activate neurons within the amygdala to signal other
brain regions to strengthen memory for what is happening (ibid.: 20–1).

The central and medial areas of the amygdala are connected
to the hypothalamus, which is where the hormones needed to
trigger fight-or-flight actions are secreted. Linkages to the
corpus striatum and brain areas that monitor movement are
found in the basal area. Via their connections to the central
nucleus, the amygdala can also generate cardiovascular, muscular, and gut reactions of the autonomic nervous system.
Finally, the amygdala is linked to the brain stem locus coeruleus,

an area where norepinephrine is manufactured, to be distributed to the cortex, the brain stem, and the limbic system for greater reactivity to sensory intake and readiness for action.

In a situation of fear, the amygdala looks for potential threats and sends an alarm message to all other areas of the brain. It commands the brain stem to monitor facial expressions of fright; slow down the person's breathing; freeze unrelated movements that are already underway; accelerate the heart beat; increase the blood pressure; and activate the movement centres, preparing the muscles for appropriate action. The amygdala (together with the hippocampus) also triggers the secretion of fight-or-flight hormones and the release of norepinephrine and dopamine, key neurotransmitters that enhance brain reactivity. An alarm signal is thus sent to the senses, commanding them to focus all their attention on stimuli relevant to the emergency at hand.

While this happens, cortical memory systems are sifted so that knowledge and memories of similar stimuli can be tapped, leaving aside all thoughts that are irrelevant to the immediate situation (ibid.: 17, 299). The hippocampus plays a central role here. Thanks to its action, "the very same neurochemical alerting systems that prime the body to react to life-threatening emergencies by fighting or fleeing also stamp the moment in memory with vividness" (ibid.: 20). If the hippocampus confirms that there is reason to panic (an object resembling a branch or a snake could be a snake after all); the prefrontal lobes increase and fixate the attention on incoming stimuli, looking for more information. If still without reassurances, the amygdala sends an alarm that triggers the hypothalamus, the brain stem, and the autonomic nervous system into action. The alarm signal is tailored to the situation, activating neurons equipped with receptors primed for particular neurotransmitters.

Prefrontal lobes (anterior to the motor cortex) are in charge of evaluating and directing attention to stimuli processed in response to signals received from the amygdala and the hippocampus. While warning signals come from the limbic system, the application of attentionality and judgment to a threatening

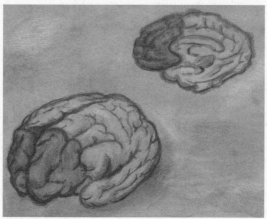

The prefrontal cortex. Drawing by Martin Blanchet

situation requires that impulsive reactions of limbic origin be kept in check, another function of the prefrontal brain. Provided that the amygdala does not highjack the brain into emergency action, connections to the anterior part of the frontal cortex (as distinct from the posterior motor and premotor regions) allow the brain to monitor or stifle intense emotions such as fear or anger. Prefrontal lobes thus weigh the adequacy of responses to our impulses, keeping emotional reactions of the limbic system within bounds. This is the most complex area of the cortex. It is distinctive of the human species in that it allows subjects to control unpleasant emotions and impulsive responses to emergency situations. As Goleman (1995: 26) puts it, while "the amygdala proposes, the prefrontal lobe disposes." The author adds that the left lobe is particularly important in this regard: "the right prefrontal lobes are a seat of negative feelings like fear and aggression, while the left lobes keep those raw emotions in check, probably by inhibiting the right lobes." More shall be said on these hemispheric differences in our review of the "autonomic lateralization" literature.

While they are responsive to emotions and derive motivation from subcortical connections, prefrontal lobes keep emotive connections in check, generating moral and instrumental behaviour through limbic inhibition. The prefrontal brain thus takes the lead in activities involving ethical conduct, step-by-step planning, and instrumental action. It introduces logical schemes, normative rules, and practical reasoning into pursuits otherwise grounded in limbic motivation. The logical and the acceptable are brought together, as when the subject carries out plans and "abides by the rules." Behaviour or propositions that reflect conventional thought, moral conduct, and goal-oriented action are thus dependent on prefrontal input. As Heilman (1995: 224) remarks, "unlike limbic system input, the frontal input into the attentional systems may provide information about goals that are not motivated by an immediate biological need." Altruism, sociability, and normal personality are severely affected when the prefrontal lobes are affected by the consumption of alcohol. They also suffer when part of the lobes are removed or severed from the lower brain, as in lobotomy experiments of earlier decades.

Subjects with lesions to the frontal lobes show three sets of symptoms that confirm prefrontal involvement in the attentions of normative and instrumental behaviour. The first syndrome is associated with insults to the orbitofrontal area and generates signs of disinhibition such as impulsive sexuality and asocial behaviour. The second syndrome is one of apathy and apparently affects the mediofrontal area. Symptoms include not only persistent apathy (with bouts of fits) but also reduced sensitivity to pain, lack of spontaneity, shallow emotions, mutism, and diminished motoractivity (e.g., facial expressions, eye movements). The last syndrome is of a "dysexecutive" nature and appears to be caused by dorsolateral injuries. It impairs the subject's capacity for step-by-step, goal-directed behaviour, resulting in indifference to past and future, lack of planning and foresight, attention deficit, and cognitive impersistence. Diminished judgment and motor programming problems such as aphasia or apraxia are also observed. Given their inability to plan and adjust their behaviour according to goal and circumstance, patients tend to imitate other people's behaviour and persevere with actions even when they are no longer appropriate (Rosenzweig et al. 1999: 547–8).

Such actions point to the goal-oriented aspects of prefrontal brain activity and the executive control mechanisms of attentionality. These mechanims include both inhibitory and facilitatory functions. They enable the brain to voluntarily stop or resume sensory and cognitive information processing and to pursue activities in the face of multiple, competing distractions. The prefrontal cortex thus plays a key role in selecting stimuli and memories to be attended to, as well as the intensity and duration of the attentions thus granted. Executive attentional control is vital to a variety of complex tasks such as planning, error detection, processing of novel or difficult stimuli, directing and coordinating multiple responses, or conflict management (LaBerge 1995: 220–1; Parasuraman 1998: 3; Posner and DiGirolamo 1998: 401; Swanson et al. 1998: 447). Given these functions, prefrontal lesions may result in an inability to inhibit distracting information; attention deficits involving hyperactive and impulsive behaviour; and noisy mental processes that impair the maintenance of vigilance and working memory systems (Swanson et al. 1998: 455; Swick and Knight 1998: 155–6). Similar attentional deficits are found amongst schizophrenic patients. Studies of schizophrenia reveal faulty inhibition and abnormal facilitation problems affecting voluntary operations and higher-order activities such as semantic processing. Indications are that fast-acting, involuntary inhibitory mechanisms, such as those found in basic sensory processing mechanisms, are also impaired (Nestor and O'Donnell 1998: 527).

Last but not least, prefrontal brain activity is needed to exercise interpretive judgment. The selective property of prefrontal attentionality makes it indeed possible for appropriate and rapid judgments to be made regarding objects and attributes in the environment. Interpretive attentionality thus acts as a gate for salient information flow in the brain (Cohen et al. 1993: 3; LaBerge 1995: 218–19).

IMPLICIT LEARNING

Forms of attentionality generated by the brain hinge on four sets of conditions. What we call "consciousness" is a ratio-based

phenomenon that responds to (1) the overall summation of depolarizing, polarizing, and hyperpolarizing potentials; (2) the activation or disactivation of the reticular formation; (3) signals of
alarm, appraisal, and autonomic regulation generated by the
thalamus, the hypothalamus, the hippocampus, and the amygdala;
and (4) the judgments, inhibitions, moral evaluations, step-by-step
actions, and goal-oriented attentions of the prefrontal brain.
These conditions mapped along neural lines and the cortical/
subcortical axis point to the complexity of brain attentional
mechanisms and their irreducibility to half-brain talk and
related specializations mapped along the sagittal plane. They
also indicate how neural activity that never reaches "consciousness" is vital to brain attentionality. Exercises in awareness
presuppose the inhibitory works of polarization, hyperpolarization, reticular selectivity, and prefrontal restraint.

The relationship between cortical and subcortical processes
tends to be asymmetrical. On the one hand, the cortex requires
subcortical support in order to activate itself. For instance, for
external stimuli to be consciously appraised, the reticular
system must be turned on. On the other hand, subcortical functions may have a life of their own, even when monitored or
kept in check by the cortex. Amygdala activity is a case in
point. As already noted, studies of what the amygdala does in
emergency situations show how some parts of the brain must
occasionally bypass cortical involvement and proceed directly
to an autonomic processing of stimuli and emotions at hand.

Another instance of ratio-based attentionality is *implicit
learning*. Not all forms of information processing require full
attentionality. Actually, information acquired selectively and
through uneven attentional levels is essential to any normal processing system. A multitude of sensory and memory input processed simultaneously and with full awareness would simply
overwhelm the system's computational capabilities (Niebur
1998: 180). Important visual information can thus be obtained
without visual monitoring and the corresponding movement of
the eyes. The information is captured but does not enter the
short-term memory or remain there long enough to reach awareness (Niebur 1998: 164). This phenomenon is known as "covert

attention." It can be summoned automatically, triggered by spatial cues presented in the periphery, or elicited cognitively, by means of central symbolic cues. Advance cues will then direct the subject to attend to a given stimulus while fixating on something else (Robbins 1998: 203; Parasumaran and Greenwood 1998: 464, 480). These studies suggest that some preattentive processing or visual stimulus categorization occurs prior to acts of full attentional selection. Implicit information processing allows the visual system to go beyond a simple spotlight strategy – a "blind application of a tunnel vision to successive locations, in a desperate search for anything 'out there'" (Driver and Baylis 1998: 300). Note that by the age of four months infants are able to perform these covert shifts of visual attention. These abilities will then further develop in terms of greater speed, flexibility (changing field size), and disengagement capacity (Johnson 1998: 427, 440).

Implicit learning occurs at the cortical level, but it may also implicate subcortical functions that exercise relative autonomy and generate processes essential to "higher" brain activity. The notion that implicit learning can be achieved through subcortical mechanisms is well illustrated in split-brain studies of interhemispheric communications. Liederman (1995) argues that an important role of subcortical pathways is to allow unconscious access to information and modes of inquiry distributed in various regions of the brain. Her claim is that disconnection problems observed amongst split-brain patients affect the integration of explicit information across regions but not the subject's access to implicit knowledge processed and stored at subcortical levels. For instance, patients may not be able to name (LH output) the Queen of England shown to the right brain in a picture, yet they can speak about the queenly attributes of the person. The patients implicitly know things about a person they cannot verbally identify. Similarly, when numbers are shown separately to each hemisphere, they cannot say if the numbers are identical, but they can compare them in other ways (greater or smaller, adding up to odd or even numbers and to more or less than ten). Finally, they cannot identify the word formed by letters received by each hemisphere, yet they are able to tell whether or not the letters constitute a word.

How should we explain this incomplete information processing observed amongst split-brain subjects? According to Liederman, the answer lies in the asynchronic and fragmented nature of the information transferred between the two hemispheres via subcortical pathways. Since the two hemispheres of split-brain subjects do not work in synchrony, they can no longer produce explicit, fully formed event memories. Without a callosum, they cannot integrate the combined input of the two hemispheres by means of simultaneous processing; detailed information about item-specific identities cannot be consciously attended by the two brains (Liederman 1995: 482–3). Some information can nonetheless be channelled through subcortical circuits. The transfer is done in ways that are "fragmented and asynchronized and will not permit the various dimensions of stimulus encoding to be transmitted together as a single event. Therefore, the available memory will not have been tagged with the idiosyncratic markers of emotion, timing, history, and so forth, that make it retrievable as a singular, episodic, experience bathed in its contextual surround" (ibid.: 462). A name cannot be attached to the picture of the Queen of England shown to the RH since the picture does not congeal into a full episodic event memory, with complete temporal and autobiographic elements. Despite this limitation, some categorical or attributional fragments are transmitted to the LH through subcortical pathways. The patient can say that the unknown person appearing on the picture is of high rank.

A similar form of implicit learning occurs when one hemisphere receives information and uses it in a fragmentary fashion, without full attentional integration. In the case of split-brain patients, insufficient arousal and activation of one hemisphere result in a one-sided internal neglect of the information received. External information is not fully processed by the hemisphere that suffers a weakening of internal integrative activity due to callosal rupture. For instance, if the LH is weakened due to callosal section, mutism may result. Speech problems are all the more likely to happen if one hemisphere controls the dominant hand but not speech; in the absence of callosal coordination, this hemispheric division of labour creates a compulsion to activate one brain at the expense of the other. "After callosal section, the underactivated hemisphere is rendered inattentive in a manner similar to that which occurs after a right parietal lesion. From this perspective, the corpus callosum is critical for regulating the distribution of activation between the hemispheres" (Liederman 1995: 468).

Under normal conditions, however, some imbalance between the two hemispheres may still be required and attained through callosal mechanisms. The main function of the callosum is not to transfer information but rather to equilibrate the simultaneous activation of the two hemispheres through subcortical brain stem mediation. This equilibrium function may involve increasing arousal in one hemisphere while reducing it in the other (by inhibiting dopamine release), according to the task at hand. As Zaidel (1995: 491) points out, reciprocal inhibitory activity operating through the callosum involves "mechanisms of control that maintain functional independence, initiate and stop exchange of information through the corpus callosum, resolve conflict, and establish priority" (see Liederman 1995: 466).

Normally this synchronic equilibrating mechanism will not impede each hemisphere from doing some internal integration. What happens when explicit information is lost by split-brain patients is that both hemispheres succeed in receiving or detecting external stimuli, but one of them fails to integrate the information. Some of the information is nonetheless retained. Callosum-lesioned patients will not be able to name the difference they observe between two pictures of houses that are identical, except that the one *on the left side* (RH) is burning. However, if asked in which house they would prefer to live, they will predictably reject the house on fire. Likewise, a RH patient may not be able to recognize a familiar face but may show a positive electrodermal response to the same face. It is as if the subject knows something about the face but can not pay attention to the full biographic data attached to it. Chimeric tests (figures with two different facial halves fused into one) applied to split-brain patients point to similar conclusions. They show that bilateral information is successfully encoded by the two hemispheres, but that identifications and stimulus closure or completion are made on the basis of which hemisphere is being depressed and which is *consciously activated*. The LH (processing right visual field information) takes the lead if a verbal description is solicited ("name the face" commands). Conversely, the RH (processing left visual field information) takes the lead if identification is made by means of pointing or picking. Again, these experiments suggest that subjects can know or experience something "implicitly," without signs of explicit awareness (Churchland 1986: 224, 227).

In short,

disconnection symptoms, such as those that are seen in split-brain patients, can be accounted for mainly by two factors: the desynchronized and fragmented manner by which subcortical pathways permit interhemispheric integration, and the diminished arousal state of the nonviewing hemisphere without the synchronizing influence of the cerebral commissures. The under-activated hemisphere displays an "internal neglect" that is marked by an abnormality of processing of input after sensory reception; it does not effectively process inputs to it (a) directly from the contralateral side of space or (b) indirectly from the opposite hemisphere. These underprocessed inputs are not consciously perceived though they are reacted to on an implicit level (Liederman 1995: 483).

Readers should bear in mind that intra-hemispheric fibres and processing operations are more numerous compared to cross-hemispheric connections. Nonetheless, cross-hemispheric activity is better suited to some integrative tasks such as conscious identity information, a task that is "vulnerable to regional differences in arousal or emotional state, which are greater inter- than intrahemispherically" (ibid.). Split-brain patients may compensate the absence of integrative pathways of the corpus callosum by resorting to subcortical circuits and some cross-cueing as well, especially in situations involving simple binary choices. But the information thus transferred from one hemisphere to another is incomplete and tends to be categorical, connotative, implicit-automatic, and affective.

A basic conclusion that follows from studies of frontal lobe activity, limbic knowledge, and implicit learning is that *attentionality must be analogical in the sense of being ratio based by necessity*. While high levels of attention are essential to normal brain activity, "there is now evidence that not all attentional processes require awareness" (Cohen et al. 1993: 7). At the cognitive level, the attentional algorithm is based on a positive difference between sensory and mnemonic information flow at the attended site and the information flow in its surround. The strength of this difference varies in intensity, from levels that summon limited attention to levels that can "possess" or "fill" the mind. The attentional result will also vary in duration, from relatively brief or transient moments to relatively long and sustained periods of time (LaBerge 1995: 217, 221–2).

But the algorithm should include subcortical operations as well. Gradients of attention thus range from autonomic acts of limbic assessment (as in amygdala activity) to implicit learning involving incomplete subcortical processing and lower levels of hemispheric arousal and integration. When fully deployed, quanta of "organic consciousness" add up to the complex attentions of full cognition, instrumental reasoning, logical planning, interpretive judgment, and moral assessment. This attentional calculus rests upon massive sums of synaptic inattentions (polarizations, hyperpolarizations) as well as measures of reticular selectivity and prefrontal restraint. Energy expended on non-communications between neurons and also between cortical and subcortical areas is essential to the living brain.

Autonomic Lateralization

Not all neural activity requires or leads to cortical awareness. Unaware processing is in fact the rule rather than the exception. It is so fundamental as to be necessary for the brain to function. It also constitutes a manifold activity in its own right. Unaware "conversations" take several forms, including peripheral autonomic communications; synaptic depolarizations achieved without neocortical processing or reticular arousal; implicit learning involving subcortical pathways; and limbic activity monitored (channelled or censored) by the prefrontal brain. We have seen that brain attentionality is also the tip of a massive iceberg of polarizations and hyperpolarizations, billions of synaptic actions of noncommunication that go literally unnoticed. Neurons can interconnect and generate awareness on condition that energy is put into blocking off useless and inappropriate communications that would impede the brain from doing what it is designed to do.

Subcortical operations not only create the subterranean conditions and grounds for processes carried out by the two cortical hemispheres. They also fashion the emotive profile of each hemisphere. This brings us to the cortical and subcortical structures involved in connecting hemispheres with emotive valence (Lane and Jennings 1995: 294–5). This issue concerns not so much the cerebral processing of *information about perceived emotions*. What is at stake rather is the *emotive experiences or dispositions* exhibited by each side of the brain (Davidson 1995: 364). This requires that we explore the emotive and physiological aspects of neurological asymmetry, autonomic effects that may not be directly reflected in conscious activity. But first we look at different characterizations of right-brain and left-brain affects and feelings.

Using electrophysiologic measurements of cortical activation, Davidson (1995) proposes a model whereby the left frontal region is the locus of intentional behaviour and planning. The LH is thus characterized by a "wilful approach" affectivity, a positive attitude fostered in potentially rewarding situations. Note that the LH's voluntaristic profile tallies with the importance of right handedness in the development of the human organism.

When approach-related affectivity is inhibited, the cortical and subcortical regions of the right frontal brain are left to themselves, with their propensity to negative withdrawal reactivity, including reactions of fear and disgust associated with potential punishment situations. When left unchecked, negative affects may become pathological and lead to anxiety and behavioural inhibition. Symptoms include apathy, psychomotor retardation, depression, low energy, indifference to pleasure, and loss of interest in others. Similar symptoms are associated with a left anterior inactivation and lesions to the left frontal brain (see also Bruder 1995: 676). Davidson (1995: 377) adds that subjects displaying a right-frontal affective style "show significantly less natural killer cell activity compared with the left frontal subjects," with the implication that they have weaker immune systems.

In a similar vein, Goldstein and Gainotti report that lesions to the LH produce catastrophic reactions (tears, despair, anger); by contrast, lesions to the RH are associated with symptoms of indifference (euphoria, lack of concern, unawareness). Older experiments of hemispheric sedation with amobarbital sodium led to similar conclusions. That is, the left side tends towards positive affects, and the right towards negative mood reactions. Right hemispheric activity unchecked by the other brain (due to a LH lesion) may therefore cause pathological crying (instead of the pathological laughing that may result from a RH injury; see Liotti and Tucker 1995: 390).

The LH is inclined to take flight (or to fight) when need be, an attitude that may seem incompatible with the left-brain tendency to approach the stimulus or seize the object of its attention. Flight action appears to be more in line with the RH inclination to put things into perspective, withdrawing or taking a distance from things perceived by the senses. But the inconsistency is more apparent than real. Fleeing out of fear is not the same as withdrawing and standing back (from the trees to see the broader forest). Far from implying cognitive and emotional distance from the object of one's fear, the action of fleeing presupposes focalization – actively focusing on the object of fear and means to avoid it.

As already suggested, injuries to the right or left sides of the frontal areas generally confirm the emotive differences observed between the two hemispheres. The frontal lobes, however, are not the only areas known to have an impact on emotional states when damaged. Subcortical and posterior regions of the brain must also be considered. Some studies thus suggest that affects or tendencies "let loose" on one side result from *disconnections between same-side cortical and subcortical functions*, not from lack of input from the other hemisphere. That is, affective disorders are not caused by one cortical region being "released" from another due to impairment. Rather they are caused by damage affecting ipsilateral connections along the vertical dimension or axial plane. In keeping with this hypothesis, subcortical lesions to the thalamus, the anterior caudate, and the basal ganglia have been observed amongst mania patients. Thus "many manic symptoms, such as euphoria, hyperactivity, and insomnia, may be explained by a disruption of the control exerted by orbitofrontal cortex over septal, hypothalamic, and mesencephalic regions (Nauta 1971). In agreement with a vertical model of ipsilateral release, mania may be triggered by disinhibition of intact subcortical centers as a result of cortical dysfunction" (ibid.: 396–7).[1]

Similarly, Robinson and Downhill (1995: 708)[2] note that biochemical changes associated with brain lesions, particularly those involving serotonin and norepinephrine, may have an impact on clinical disorders and related differences in hemispheric behaviour. But how can chemicals regulated through autonomic pathways impact on hemispheric asymmetry phenomena?

Before we answer this question, a few words should be said about the *autonomic nervous system* and its bifurcation into sympathetic and parasympathetic pathways. Briefly, the *parasympathetic system* consists of motor nerve fibres that originate in the head and sacral regions of the spinal cord and that maintain muscle tone, induce secretion, and dilate blood vessels. This system basically takes care of routine activities of the organism, at least while it is relaxed. By contrast, the *sympathetic system* has motor nerve fibres that originate in the cervical, thoracic, and lumbar regions of the spinal cord and serve to depress glandular secretion, decrease muscle tone, and induce blood-vessel contraction. It mobilizes the organism in situations of physical exercise, stimulating it into muscular and skeletal action. When this happens, digestive and urinary activity slows down, blood vessels and viscera contract, and the heart, bronchioles, and skeleton muscles dilate (increasing blood and oxygen intake). The sympathetic system is at work in extreme situations of flight or fight. Related feelings of fear or anger produce symptoms such as deep and quick breathing, faster heart beat, perspiration, goose pimples, cold hands, and higher blood pressure (resulting from blood vessel contraction).

Both branches of the autonomic nervous system involve nonvoluntary pathways innervating the smooth muscles, the heart muscle, and the glands. Most of the signals they channel do not reach consciousness. Autonomic pathways thus differ from the *somatic, voluntary nervous system*, consisting of motor nerve fibres that send messages to skeleton muscles. When combined, voluntary and non-voluntary systems form the *peripheral nervous system* (PNS) consisting of pathways to and from the *central nervous system* (CNS). The CNS comprises the brain and the spinal cord; it transmits motor impulses while also receiving and interpreting sensory impulses.

Autonomic conversations channelled via sympathetic and parasympathetic pathways involve depolarizations performed

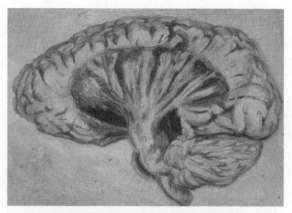

Descending fibres. Drawing by Martin Blanchet

without hemispheric awareness and cortical attentionality. They proceed through an axial mapping of neural activities, as distinct from sagittal connections between the two hemispheres. But the two sets of pathways, vertical and lateral, cannot be separated, and they are known to interact. Thus all indications are that autonomic factors are responsible for emotive differences that lie between the two hemispheres. In the words of Wittling (1995: 335), hemispheric asymmetry characterizes "not only cerebral control of cognitive functions or subjectively experienced emotional states but cerebral regulation of autonomic-physiologic processes as well." Lateral brain asymmetry is affected by the whole neural system. This is not surprising, given that the lateralization phenomenon operates at all levels of the nervous system, including not only the lower brain stem but also the midbrain, the hypothalamus, the thalamus, the basal ganglia, and the amygdala.

Vertical pathways engaged in autonomic control follow same-side (ipsilateral) routes. These top-down connections suggest that "higher" cognitive and emotive-limbic processes are coordinated with autonomic functions and homeostatic needs of the body. Cortical zones such as the medial prefrontal cortex, the cingulate gyrus, and the insula have ipsilateral connections with amygdala, hypothalamus, brain stem, and spinal cord areas.

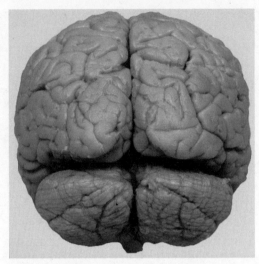

A rear view of the two hemispheres. Sundsten,
The Digital Anatomist.

The prefrontal cortex and the orbitofrontal regions play a particularly important role in the cerebral regulation of physiologic
functions. The amygdala is in turn hooked up to the hypothalamus, the thalamus, the midbrain, the pons, and the medulla.
To these same-side connections can be added direct linkages
between hypothalamic areas and preganglionic nuclei of the
sympathetic and parasympathetic systems (ibid.: 340–1; Lane
and Jennings 1995: 279).

Physiologic functions laterally modulated by the cortex
include neurotransmitter and cardiovascular activities, pain
sensitivity, immunologic responsiveness, and neuroendocrine
processes.[3] The biochemical aspects of brain lateralization can
be observed through asymmetries in concentrations of neu-
rotransmitting catecholamines. The chemicals in question originate from the medulla of the adrenal gland and affect the
sympathetic nervous system, producing arousal in situations of
fight or flight. One such substance is *dopamine*, with the left
brain showing greater quantities of it. This neurotransmitter is
particularly important in initiating and regulating movement

and motor skills, as in speech. It is mostly found in prefrontal, premotor, and motor areas. Dopamine is "involved in higher integrative cortical functions and in the regulation of cortical output activity, especially in motor control" (Wittling 1995: 309). It also permits the tonic activation of a motor readiness system involving focal or local attention efforts (Liotti and Tucker 1995: 401). An insufficient quantity of dopamine in the substantia nigra area is found among subjects suffering from Parkinson's disease.

By contrast, the right hemisphere and right-side thalamus have greater quantities of *norepinephrine*. This hormone is secreted in the area (locus coeruleus) located within the reticular formation which filters and passes on sensory information to the cortex. Norepinephrine can either excite or inhibit and will act on the cortex, the thalamus, the hypothalamus, the cerebellum, and the spinal cord. While an excess of norepinephrine may cause stress or neurosis, lack of it may lead to depression. The substance is thus used in antidepressant drugs. Norepinephrine can generate feelings of pleasure, but it is also responsible for the contraction of blood vessels, which is typically associated with a faster heartbeat and with situations of stress or emergency. Note that norepinephrine-related states of fear or anger felt predominantly in the RH may trigger the LH and the organism into dopamine-regulated motor action and related strategies of flight or fight.

The RH (especially the prefrontal cortex) also plays a leading role in monitoring levels of arousal (together with the reticular formation, the thalamic nuclei, and the basal forebrain; see Parasuraman et al. 1998: 221, 235–6). Two important chemical substances are involved here. The RH contains greater quantities of norepinephrine and shows a higher level of noradrenergic innervation, a biochemical process causing arousal. However, *serotonin*, the neurotransmitter that reduces arousal and noradrenergic activity, is also found in greater concentrations in the right brain, at least in normal nondepressive subjects. (Suicide victims show a reversed lateralization.) Thus "both neurotransmitter substances being mostly involved in the upregulation (norepinephrine) and downregulation (serotonin) of arousal, and having reciprocally inhibitory relations ... are asymmetrically distributed, favoring the right side of the brain" (Wittling 1995: 309). Imbalanced combinations of norepinephrine and serotonin can produce pathological symptoms. High phasic arousal problems may result in mania symptoms, whereas low phasic arousal is associated with depression (Liotti and Tucker 1995: 402).

Brain asymmetry affects the hormonal information circulating through the blood. The *hormonal system* is controlled by the hypothalamus and therefore the midbrain, the limbic system, the amygdala, and also the frontal cortex. There is chemical evidence pointing to a greater right-brain involvement in these vital functions of the body. *Cortisol* secretion is a case in point. While chronically elevated cortisol levels and an excess of defense reactions may increase vulnerability to diseases, adequate levels enhance resistance to problems of immunoreactivity and stress (Wittling 1995: 315). More of this substance is found in the right hemisphere. One experiment confirming this involves projecting films with variable emotional valence to one hemisphere at a time. The experiment shows a neuroendocrinological reaction in the RH only, with an increase in levels of cortisol (ibid.: 312). Yet cerebral control over this hormonal activity is not a conscious process. When subjects were asked questions regarding their emotional responses to these films, their responses showed no difference between film presentations processed through right and left visual fields. The implication here is that asymmetrical neuroendocrine regulation acts independently of subjective emotions associated with RH arousal.

These findings on neuroendocrine secretion and neurotransmitter activity involved in arousal suggest in turn that our *immunological system* responds to lateral differences in cerebral structure. In the words of Wittling (ibid.: 318), "since there is widespread evidence that lateralization is a main characteristic of anatomic and functional brain organization, and that several mechanisms involved in brain immunomodulation, such as emotional arousal, sympathetic innervation, neurotransmitter concentration, and neuroendocrine activity, are clearly lateralized, it seems most likely that the brain modulates immune response in an *asymmetric* manner."

Studies of *cardiac activity* also show a tendency towards right brain specialization. Whereas the ability of the heart muscle to contract is better controlled through left-side sympathetic pathways, studies of cortical and subcortical structures modulating the heart rate reveal a predominance of sympathetic and parasympathetic pathways passing on the right side (originating primarily in the hypothalamus and the amygdala). Stimulation of the right cranial vagus will thus cause significant changes in the heart rate (deceleration) compared to left vagal nerve stimulation. Given this asymmetrical structure, cardiac arrhythmia can be caused by a blockage or an ablation of the right stellate ganglion. Likewise, parasympathetic control of the heart rate is more seriously affected by damages to the right brain compared to left brain lesions.

In keeping with these findings, faster heart rates have been observed amongst subjects anticipating emotionally charged slides presented to the RH (hence appearing in the left hemifield). Increases in heart rate and blood pressure (especially amongst female subjects) can also be obtained through RH film presentations of disturbing slide or film material. While sensitive to negative affects, the RH's predominant role in regulating autonomic functions means that this hemisphere is likely to show greater vulnerability to pain, a sensitivity extending to the left side of the body. Compared to normal subjects and left-brain patients, patients with lesions to the RH thus show greater endurance to pain. They are also less facially expressive. As well, when subjected to disturbing film material or emotionally charged stimulation, they show reduced affective and autonomic responsiveness (skin conductance, heart rate). Other studies suggest that patients with RH damage (and epileptic subjects with right temporal foci) experience more frequent sexual dysfunction and lower levels of arousal compared to normal subjects (ibid.: 323–7, 331–5; Hugdahl 1995: 241; Lane and Jennings 1995: 293–4; Liotti and Tucker 1995: 391).

The RH is apparently better at assessing changes in cardio-vascular activity, thereby confirming the preceding observations. Still, left-right differences in heart-rate control and blood pressure are not necessarily reflected in subjective assessments of cognitive or emotional arousal. Again, the asymmetrical regulation of autonomic-physiologic functions appears to be independent of hemispheric variations in cognitive and emotional attentionalities. This can be verified in studies of postmortem variations in hemispheric neurotransmitter concentrations, studies where arousal effects are eliminated.

To sum up, "while dopaminergic activity is predominately left-lateralized as is neural control of AV [atrioventricular] conduction and myocardial contractility, all the other functions are predominantly right-lateralized, including noradrenergic and serotonergic neurotransmitter activity, most neuroendocrine activity, chronotropic cardiac activity, arterial blood pressure, cardiac feedback and discrimination, skin conductance response, and pain sensitivity" (Wittling 1995: 336). The right brain thus appears to have a closer connection to limbic circuits (Liotti and Tucker 1995: 394). It plays a leading role in the control of vital functions involving survival of the organism and its constant adaptation to stress and external challenges.

Brain asymmetry at the autonomic-physiologic level is so critical to the human organism that variations or reversals of normal lateralization patterns can lead to a greater frequency of physical disorders. More abnormalities of the affective or behavioural kind and higher levels of physical complaints are thus associated with inversions in levels of serotonergic activity, cortisol secretion, or event-related potentials evoked by emotional stimuli. Likewise, if normal patterns of cardiac regulation are altered, cardiac arrhythmias, ECG abnormalities and sudden death can occur. Lastly, there is evidence to suggest that left-handedness may result in lower levels of cortisol secretion during stress-related situations. This can lead to greater frequencies of immunologic and other physiologic disorders and reduced longevity as well (Wittling 1995: 343).

In conclusion, a multitude of signals transmitted through the nervous system never reach central nervous consciousness. Some neural conversations are given cortical attention and may result in voluntary somatic events (muscular and skeletal) channelled through the peripheral nervous system. Thousands of synaptic conversations, however, are performed without awareness or voluntary control. Involuntary impulses and neurochemical substances are constantly transmitted through peripheral autonomic pathways. These communications take on sympathetic or parasympathetic functions that automatically stimulate or inhibit nervous activity (such as fight or flight), respectively. Autonomic activities also create emotive dispositions within each hemisphere. The left side thus exhibits a positive-approach attitude and shows greater concentrations of dopamine supporting heart-muscle contraction and motor-control activity. By contrast, the right side displays a negative-withdrawal reactivity sustained by norepinephrine and serotonin, neurotransmitters involved in the upregulation and downregulation of arousal. Also, the RH is more directly involved in maintaining heart rate, arterial blood pressure, sensitivity to pain, and hormonal activities serving neuroendocrinological and immunoreactive functions.

Autonomic processes generate communications without the cortex showing awareness of them. The central nervous system can be inattentive to its own activity, a phenomenon that may result in implicit learning. Subcortical processes can generate "unaware learning" that can be exchanged between hemispheres through brain stem mediation, as shown in studies of split-brain subjects. Also, signals of limbic origin – for instance, those channelled from the thalamus to the autonomic system via the amygdala – may trigger reactions that do not fall under the control of prefrontal attentionality and goal-oriented activity.

Readers should recall that unawareness is not merely a passive phenomenon – the neocortex simply ignoring communications that are in no need of full brain arousal. More to the point, the central nervous system puts considerable energy into blocking the route to inadvisable connections. This means making sure that some communications do not happen, while others do, but without the cortex having to take notice of them. These effects can be obtained through inhibitory mechanisms of the corpus callosum, interhemispheric pathways in charge of synchronizing the two hemispheres but also granting priority to one lateral specialization over another. Active inhibition can also be obtained through prefrontal checks on limbic activity. Finally, it can be achieved through selective mechanisms of the reticular formation. This formation awakens the brain to external stimuli and then filters those that should not be attended lest the system crash down through sheer exhaustion or information overload.

Inattention effects exist at the neural level as well. The prospects of synaptic pathways reaching central nervous attention can be restrained in one of three ways. First, while facilitating impulse transmission, potentials may be of the graded kind, providing information without triggering cells and axons into action. Second, synaptic zones can be instructed to remain polarized and at rest, thereby maintaining capacitance and withstanding impulse transmission. Lastly, synapses can be instructed to hyperpolarize (increasing intracellular negativity) and disconnect while other neurons depolarize (decreasing intracellular negativity) and communicate.

SEMIOTIC ATTENTIONS

The Rank Ordering of Signs

Neurons form webs and circuits of synaptic communications. Some neural conversations arouse attention, triggering the reticular system into action and letting the brain become aware of some depolarizations and action potentials. Great numbers of impulses, however, are transmitted without our knowing it. Thus a basic feature of the neural process is that the two forms of synaptic exchange, the attentive and the inattentive, do not always intersect. While the two are vital to integrated brain activity, warps of neural connectivity do not always intersect with woofs of attentional noticing. Not all things going on inside our brain need occupy our mind, so to speak. In point of fact a lot of brain chemistry and electricity goes into avoiding countless connections and simply "minding" our business, not concerning ourselves with things that can be done automatically. In the end, cortical awareness presupposes massive sums of neural inattentions. Operations that come out into the open rest on mountains of chemical and electrical closures, shutting off activities ranging from resting potentials and cellular hyperpolarizations to autonomic, reticular, limbic, callosal, and prefrontal checks on brain attentionality.

Sign networking works in the same way: attentions and inhibitions also play themselves out in semiosis. To paraphrase the language of neuropsychology, the nervous sign system comprises four layers of activity: the somatic, the autonomic, the hyperpolarized, and the polarized.

First, semiosis generates *somatic* sign communications. These consist of action potentials noticeable in language, depolarized sign impulses that constitute the overt attentions and expressions of semiosis.

Second, not all sign interconnections are effectively noticed, occupying a proximal position vis-à-vis centres of awareness. The majority of conversations are rather of the distal sort – connections activated through *autonomic* pathways, circuits located at a distance from the explicit and the overt. Autonomic exchanges consist of graded implications that facilitate the process of communication, supportive pathways that produce the subsidiary impressions and evocations of nervous sign activity.

Third, inhibitions applied to unwanted connections must also be factored in. They consist of *hyperpolarized* interspaces, sign pathways that show high levels of resistance or capacitance. As with neurons, "signaptic" hyperpolarization involves the application of force or energy against inappropriate connections, placing them in a state of great tension.

Fourth, semiosis evolves in a broader context of unattended sign pathways, connections that fail to happen for one of two reasons: they have not been learned, or they are *polarized and at rest* because unrelated to signs in current use.

In short, signs can connect through either somatic or autonomic pathways, and they can disconnect through either polarization or hyperpolarization. I should emphasize that these four possibilities are not to be understood digitally, as either/or options that are mutually exclusive. Rather, they correspond to mechanisms that constantly interact, ratio-based operations that come together in summations of semiosis and account for the uneven attentions and dynamics of sign activity. As argued below, these mechanisms are so pervasive as to affect all levels of semiotic activity: the cognitive (RH, LH), the emotive (limbic), and the normative (prefrontal).

At the cognitive level, connective and disconnective forces affect the ways in which attentionality is unevenly granted to lines of convergence and divergence. Syncretic and diacritic linkages may be made *explicit*, producing somatic sign connections, say, between the falling-fig imagery and overt expressions of retribution. But cognitive linkages can be suggested through autonomic inscriptions, thereby *implying* ties between a fig apron and the Fall that comes immediately after, to use the same imagery. Meanwhile some connections may be *polarized* and left out because simply irrelevant to the task at hand (the abbreviation "fig." is unlikely to be read into a text that speaks of a leafy apron worn by Adam and Eve). Or they may be avoided and *hyperpolarized* because diametrically opposed to expressions of the surface script. That is, the fig apron imagery of Genesis 3 must be read in ways that necessarily preclude the good life proverbially enjoyed under a fig tree.

The same can be said of emotive sign connections. They too can be rendered explicitly ("soma" affects communicated *on line*). They can be made implicit ("autonomic" mood transmitted *between the lines*, so to speak). They may be avoided because inapposite ("polarized" sentiments proceeding *along other lines*). Or they may be silenced on account of their illicit character ("hyperpolarized" feelings squarely *out of line*).

Normative connections are also subject to this fourfold equation. By norms I mean two things: signs of morality and indices of rational-instrumental activity. Both can be pursued *explicitly or implicitly*. Rules as to how things should be done with ethical or rational intent can be declared or followed *attentively* (for example, using hexametric prosody, or asserting the virtues of sacrifice). But they can also be *implied or applied mechanically*, without anyone taking notice of the norms that are brought into force (for example, automatically pronouncing the accented and unaccented syllables of line 1 in *Evangeline*, or using the fig apron motif to imply the ethics of sexual shame).

Polarization in the normative domain, however, is another possibility – for instance, ignoring rules that have some connection but no real bearing on the subject at hand (e.g., tying the fig apron imagery to precise rules on how to prepare a fig poultice). As already pointed out, the polarization of irrelevant sign connections is a relatively weak form of disconnection. It simply means that "A" precludes "not A" and that the preclusion can be achieved by keeping the action potentials of "not A" at rest (ignore the fig poultice when reading Genesis 3; don't connect "fig" with "fig."; don't use pentameters when writing hexametric poetry). Polarization requires little effort, at least in comparison to another form of disconnection that calls for the application of greater energy: hyperpolarization, that is, closure obtained through active inhibition.

The distinction between polarization and hyperpolarization has a direct bearing on our discussion of connections and disconnections pervading all levels of semiosis. Briefly, while polarization is compatible with signs of normative activity, the use of greater force of closure seems not applicable to the "rule of law." After all, how can norms prescribing how things should be done (or proscribing infractions to the rules) be banned from the explicit declarations of language and the implications thereof? Given that ignorance of the law is no excuse, provisions are usually made for the law to be known. By nature norms are seemingly *inhibitive, not inhibited.*

The notion that norms can be operative while also being repressed seems utterly contradictory. But the opposite scenario is even more illogical. That is, the notion that normative activity is always declarative and explicit produces nonsense. This is particularly true of norms understood morally. To argue that morality must spell everything out (except for things that are so obvious that they will go without saying) is to neglect the fact that silence must be observed if morality is to show some coherence. Sign restraint is particularly important when language faces the delicate task of identifying things that are so reproachable that they become unspeakable – typically signs of limbic "goodies" that offend the moral good. When passing judgment, morality cannot bring out "the truth and nothing but the truth" lest it should fall into indecency. Given this problem, the task of addressing the question of something "out of the question" is bound to be two-sided. On the one hand, the unthinkable cannot be uttered, and measures to that effect (avoiding certain utterances) must be taken. On the other hand, full silence cannot be observed if morality is to signify. Some degree of indiscretion is inevitable. Moral sign actions are bound to betray their secrets, if only by means of euphemism or displacement. When combined, these two tendencies compel morality to speak up but with caution, with great concern for concealing its "private parts," so to speak. Moral circumvolutions point to the devious pathways of a code of honour shot through with silences and infractions of its own.

The analyses presented below show how levels and equations of (in)attention govern cognitive, emotive, and normative aspects of sign activity. Imageries taken from different contexts and sources will provide examples of rank ordering effects in semiosis; interchanges between the overt and the covert; intersections between intellect, affect, and precept. (Norms understood as rational-instrumental sign actions will be taken up in *3-D Mind 3*.) Emphasis will be placed on the interplay of these multiple forces of sign processing, as opposed to a digital characterization treating sign events as either normative or emotive, expressive or repressive, and so on.

Reticle 5 contains brief illustrations of how signs of morality vie for attention, *but not too much*. Other interpretive exercises to follow show how cognitive, emotive, and moral affects lend themselves to political manipulations of all sorts and the corresponding shifts in sign attentions; battles for attention imply that semiosis is inherently malleable and political. The subtleties of sign tactics will be explored in three distinct contexts: animal representations of tensions between language groups in Canada (reticle 6); the rhetoric of ethnic identity and conflict in Africa (reticle 7); and deliberations over connections between pleasure and sacrifice, as conveyed through biblical and current-day imageries of feet, shoes, earrings, and body piercing (reticles 8 to 12).

Parenthetically, the argument concerning shifts and battles in the attentions of language will not be reduced to polity narrowly defined as conflict between groups or perspectives. Signs expressing well-established views on life must also shift if they are to attract and be worthy of our attention. Teachings of the Bible serve to illustrate this point. In the scriptures a tragic scene implying that "you will suffer if you have committed a fault" can easily shift to another that says something more edifying, as in "you can be a sacrificial hero if you accept your sufferings." This in turn may be followed by a joyful announcement, such as "you will be rewarded if you sacrifice." But the story does not end there. The script can still bring another warning to our attention: "you commit a fault if you revel in your rewards." The story can then start all over again, back to the consequences of one's fault, remedies to be explored, and the rewards thereof. As in Hollywoodland, matters of life and death, wealth and poverty, virtue and depravity, or justice and inequity can never be handled once and for all. Accordingly, sign reticles are never static. They constitute struggles for (in)attention that move people and scripts into action. The nervous sign system is a political economy of shifting considerations, attentional battles in constant movement.

Signs That Matter
and Those That Don't

Connections between signs are never just horizontal, created equal "in the mind of the subject." More often than not, they are plotted along a vertical axis, a plane designed to let some signs be elevated above others. This rank ordering is performed typically for reasons that have to do with how humans conceive a good life – which can be understood in one of the two ways, the prefrontal or the limbic. The prefrontal definition of "goodness" points to what is morally right. The limbic pertains to what is simply desirable and pleasing, whether right or wrong. Sign activity pays attention to both forms of goodness. While constructed through logic, sign activity speaks to both morality and the pleasure principle. All codes are tainted with explications and implications of the desirable and the acceptable.

Measures of goodness and related rank ordering effects are embedded in the simplest manifestations of cognitive activity. Consider elementary operations of language such as calling something by its name – hence, common nouns informed by the plain logic of similarities and differences (as expounded in *3-D Mind 1*). When seen from an analytic perspective, nouns signify by dividing themselves into lower-order differences located within higher-order classes (also divided). Thus the opposition between life and death (LH divergence) is part of a broader noun-category known as the cycle of life (RH convergence). An interesting feature of this higher-order expression consists in its euphemistic character: the notion that death is merely a moment in the cycle of life. This euphemistic effect is

obtained by letting the generic term give preferential treatment
to one lower-class term over its opposite. The cycle of life
elevates life above death, treating it as a higher-order term that
brings opposites together, without extinguishing the difference.
Through this rank ordering of signs in language, the superiority
of life over death takes on wishful and moral connotations
conveniently built into the order of things, as expressed through
classificatory logic. The lesson to be drawn from this example
is that higher-class terms do not represent shifts in referential
terminology alone. More often than not, they also propose
shifts in preferential attentionality; terms pitched at higher "levels of abstraction" grant special attentions to select terms
picked out of the rank and file.

The value we grant to life over death is inscribed in the
arborescence of signs. The act of coding thus weaves affect into
intellect. This suggests that sign reticles are cognitive, emotive,
and normative all at once; things worth attending and pursuing
are embedded in grids of logic and acts of coding.

Other illustrations of this principled logic abound. For instance,
luminosity and the one-day cycle are higher-order measurements that convey our general preference for brightness and
daytime over dimness and nighttime. Likewise, measures of
height, length, breadth, and depth both include and place themselves above their own opposites – above bodies categorized as
small, short, narrow, and shallow. So too with "degrees of
abstraction" and "levels of generality": these are expressions
that carry out a mission, which is to elevate the abstract and
the general over their own opposites, that is, the concrete and
the particular. *Hierarchy in logic points to what matters the
most.* By granting greater attention to some signs over others,
systems of classification salvage the cold rule of logic (and the
erratic dispersal of *différance*) from absolute *indifférance*.

The same rank ordering principle permits language to arbitrate tensions that pit one side of goodness against the other,
the pleasurable against the commendable. Language can do this
by inscribing the opposition and its vertical resolution inside
the same sign. The concept of "passion" can serve to illustrate

this point. The word offers multiple meanings that capture basic dualities in Christian ethics. On the one hand, it denotes an intense or violent emotion, especially sexual desire or love. On the other hand, it evokes the martyrdom of an early Christian, sufferings modelled on the death of Jesus Christ and inspired by the teachings of asceticism – abstaining from pleasures of the flesh and the good life on earth. While both definitions, the lustful and the sacrificial, are acceptable, the latter expresses ideals of moral and spiritual behaviour from a Christian perspective, to be ranked above the pursuit of carnal desires. One meaning of passion lords over the other.

These hierarchic arrangements of signs of order (cognitive) and goodness (emotive, normative) should not be confused with culture understood statically, as a value system or ideology shared by all subjects dwelling in the same surround. More realistically, *hierarchy is a battlefield of signs that vie for visibility and the attentions of speech.* Conflicts over definitions of goodness will cause classificatory grids to quarrel over many things, including what aspects of life should be voiced and what ones have lesser rights of speech. The generic "man" (or "mankind") motif used to embrace both man and woman is a case in point, one that silences the "weaker sex" at the level of species. Polemics over this sexist language have drawn attention to alternative classificatory arrangements (for example, substituting "humans" for "men"), with the implication that signs of hierarchy are debatable and never fixed in stone.

The notion of *exploitation* is another telling example of how battles over ranked meanings can leave their imprint on the conventions of language. The term can be used in two ways. Exploitation can denote the process whereby unjust profits are extracted from the work of others, as in theories of the labouring classes "exploited" by capital. But it can also evoke the process whereby alternative ends are maximized through a rational allocation of scarce resources, be they raw materials, labour power or capital. Left-wing political economists prioritize the first definition, which plays a central role in their critique of market economies governed by capital. By contrast, "formal" economists and right-wing politicians emphasize the second definition, portraying social disparities as the outcome of rational decisions founded on considerations of utility and profit. Battles over the semantics of "exploitation" have far-reaching implications. The word and related sign actions (deliberating the source of value, its equitable distribution, for example) lend themselves to moral and theoretical debates between two perspectives that are diametrically opposed. One calls attention to problems of domination and subordination, whereas the other claims the right of privilege for considerations of scarcity and rationality. With virtue and truth on its side, each theoretical pole struggles to reduce the other to silence.

Rank ordering attempts to build consensus but does not presuppose it. Moreover, it gives preferential attention to signs of (wishful) domination but grants them also the virtues of strategic silence: timeliness, brevity, and diplomacy. Although occupying key positions in language, signs of "things deemed to be good" need not always be foregrounded. In fact there are several advantages that can be derived from granting some visibility to nondominant signs. One is that imageries of tribulation or sin must be explicated for morality to eventually triumph. Morality needs to triumph in a *timely* fashion and keep quiet until then. A poem bent on telling a story of paradise lost, such as Longfellow's *Evangeline*, cannot spell out its lessons without first dwelling on the sombre signs of a people's fall, shortness of life and narrow margin of happiness (note how imageries of daylight, height, length, and breadth can be downplayed, depending on the effects that are being pursued). Likewise, stories of "sacrificial passion" do not have to blot out all the lustful evocations of "passion." Signs of morality still require that the sinful connections of "passion" be explored and made explicit, if only to consider their repercussions on the lives of the wicked.

Other benefits that signs can derive from lower levels of attentionality stem from the principle of *brevity* – saying a lot in few words and without full explications. Rules of good conduct need not always be made explicit, hence clearly explained and "made plain" in language. Some of the attention granted to what is morally good can be "autonomic" in the sense of being conveyed through implicit pathways. Some moral inscriptions may actually choose to be downright cryptic. Consider the rule according to which one should not walk under a ladder lest one should be struck by bad luck. Although simple, this popular injunction entails (1) a tacit opposition between going up the ladder and walking under it; (2) the implied binarism of good luck and bad luck; and (3) a correlation between these two oppositions. The elements add up to a full analogical proposition: good fortune is to bad luck what going up and being at the top of the ladder is to being at the bottom and walking under the ladder. The desirable and rules to attain it are signified through an abbreviated sentence amalgamating sign attentions with autonomic associations implied but never spelled out.

Rules regarding things deemed to be good can be communicated by means of implication. The effectiveness of normative language does not hinge on everything being spelled out. As with phonetics and grammar, some rulings may be operative and nonetheless go unnoticed. Readers are reminded that this form of hiding is not to be confused with operations of repression or inhibition. The rule of economy or brevity does not mean that morality must hide because punishment would otherwise ensue. It simply means that lessons can be encoded and communicated without maximum attentionality. Morality is inhibitive, not inhibited – the subject of hyperpolarization, not its object. It can express itself deviously, through metaphor and allegory, but does so merely to enhance its appeal and effectiveness.

Or so we commonly think. In reality, morality's relationship to the unconscious is as nervous and tortuous as everything else in semiosis. Notwithstanding rumours to the contrary, morality has much to gain from having secrets of its own. This brings us to another form of hiding that pervades "sign rulings" and that goes beyond the half-silences of implicit language: namely, connections that are part and parcel of the ruling itself and that must be blotted out from the surface script. *Diplomacy* and self-censorship are vital ingredients of moral sign activity.

Consider an anecdotal example, based on a personal incident dating back to the early 1980s. This trivial event took place at home and involved my four-year old daughter eating an ice-cream cone and accidentally smudging my shirt as she bumped into me. She apologized. I responded with a simple expression, "*no importa*" ("it doesn't matter"), two words which for some reason I chose to pronounce with an English accent (especially the "r" which I failed to roll). I immediately noticed the awkwardness of my reply. For one thing, my child did not understand Spanish. French was the language we spoke at home, and there was nothing apparent to justify the use of Spanish. Also, I mispronounced the words while knowing perfectly well how to pronounce them. Why should I apply an English accent to Spanish words pronounced in a French-speaking context?

As with dream material, circumstances preceding and surrounding the incident go some way in explaining the actual details and anomalies of the scene. As a professor of anthropology, I was then commuting between three linguistic environments on a daily basis: French at home, English at the university, and Spanish in my research activities and social life. This was a period during which I was experiencing difficulties at work (the details of which are of no relevance to this exercise; they too are *signs of things that do not matter*), problems that I was struggling to minimize and forget. The expression "*no importa*" pronounced with an English accent was my way of telling myself that problems connected to my English work milieu mattered as little as ice-cream on my shirt. But why the Spanish words to convey this message?

Events preceding the incident suggest a simple answer to this riddle. The night before, refugees from Salvador came to my house for dinner and narrated some of the atrocities they had witnessed and suffered in their country prior to their exile. I was shocked by what they had gone through and thought my problems were strikingly insignificant compared to theirs. Mine mattered little, to say the least. They were of no importance; *no importa* ... In a nutshell, compared to what was happening in Latin America (Spanish), problems at work (English) were trivia at home (French).

Although idiosyncratic and anecdotal, this analysis illustrates the meshing of horizontal (cognitive, i.e., syncretic-diacritic) and vertical (emotive-normative) connections in language. In its own condensed manner, my *lapsus linguae* drew cognitive parallels between differences in types of problems (domestic, occupational, international), linguistic contexts (French, English, Spanish), and levels of concern (low, medium, and high). To these cognitive correlations were added a moral imperative – do not pay attention to things that do not matter – and also an anxiety to forget what could not be forgotten. The end product is thoroughly paradoxical. By means of barely two words, the mind pays devious attention to what should be left unattended, reminding itself of what should be forgotten. The prescription built into this act of speech is therefore two-layered. It comprises a surface injunction understood literally ("who cares about a stain on my shirt!"). But it also "contains" a deeper injunction, "conveying and withholding" something that cannot be made explicit lest the mind should focus on something it wants to forget ("don't think about problems at work").

When seen from a neurosemiotic perspective, the incident revolves around sign action potentials depolarized and made explicit through "proximal" or "full somatic" expression: my daughter's ice cream on my shirt and my immediate *"no importa"* response. The scene would be unintelligible, however, were it not for the implied connections that feed into the incident. They include evocations of things domestic, occupational, and international associated with French, English, and Spanish language environments, respectively. These distal attentions are linked in turn to hyperpolarized sign potentials, inhibited pathways forming a parallel circuit closed off from efforts of attentional reticulation. Although closely tied to expressions and impressions conveyed by the *"no importa"* scene, connections that "matter the most" are repressed from the visible script. They constitute a shadow sign trajectory that matches the visible route but receives no somatic attention whatsoever. Self-censorship forms an integral part of a moral injunction that is never spelled out, at least not fully.

The *"no importa"* incident is a principled statement that judges and declares something to be unimportant. Yet the moral

it conveys is not made explicit. Also, things-that-matter-not are
not entirely expelled. The end product is thus two-sided. On
the one hand, a "bottom line" injunction is concealed beneath
the cut-and-dried prescription, inhibited by virtue of the silence
to be imposed on what should be forgotten. If anything, the
utterance implements the command to pay no attention to concerns (professional worries) inspiring the words actually uttered.
On the other hand, anxieties underlying this act of speech are
given an indirect outlet. The no-matter ruling is inevitably
defeated and contravened from the moment it is pronounced.
Through efforts of hyperpolarization, infractions of the rule
("don't think about it") make their way into conventional
statements and pronouncements of the law.

My deconstruction of the incident disposes of the incongruity
of the scene. I should emphasize, however, that my analysis
does not pretend to be an adequate "representation" or "recon-
struction" of the event and its "underlying logic." While originally hyperpolarized, connections that are depolarized through
reflexive activity no longer belong to the same sign event. Acts
of symbolic condensation and inhibition differ radically from
their interpretation. For one thing, the *ex post facto* analysis
introduces "interpretive attentions" designed to establish parallels and correlations between the event under scrutiny and
discussions of language and the brain. These hermeneutic concerns play no role in the condensed "*no importa*" incident and
represent an attentional shift in their own right. Moreover, my
analysis follows a divergent route, not the convergent trajectory
obtained through words of condensation. The analysis thus
moves away from a simple utterance. It delves into the original
event's secret underpinnings and then proceeds towards the
broader implications this exercise may have for our understanding of neurosemiosis. The exercise is done at considerable distance from home, which means that "true origins" are likely
to be lost.

Name Calling: Frogs and Beavers

A few tentative generalizations emerge from the preceding analysis. First, in lieu of offering a variety of mutually exclusive menus, sign production constantly brings together the cognitive, the normative, and the emotive dimensions of semiotic activity. Neuropsychologically +speaking, similarities and differences mapped along the sagittal plane are shot through with prefrontal rulings and limbic feelings; the logic of *synkretismos* and *diakritikos* is subject to the axial interplay of sentiment and judgment. Second, the preferences and anxieties of language are communicated through mechanisms of attentional stratification. Signs of the pleasurable and the commendable are conveyed through an uneven distribution of (in)attention; the appraisal and pursuit of things deemed to be good are achieved through the overall summation of somatic, autonomic, and inhibitory affects in language. Last but not least, moral sign actions need not always be clearly spelled out. For one thing, some normative teachings must be transmitted in a timely fashion and with an economy of words, by means of automatic implications. Also, morals are constantly faced with the delicate task of dealing with "the unspeakable," speaking to "it" while banning "it" from the surface script. Some self-censorship of morality becomes unavoidable.

The embattlement of signs vying for attention but also struggling for brevity, economy, and diplomacy applies to personal events such as the anecdotal *"no importa"* event. The same principles, however, can be extended to broader struggles in sign activity, moves and battles that may take on political proportions and affect the course of history. Mention has been made of politically loaded debates over the semantics of "man" and "exploitation." The analysis that follows looks at another type of script, one that involves popular stereotypes – more specifically, name-calling as applied to social groups.

We have seen how proper naming practices are governed by two kinds of rules. On the one hand, logical conventions generate a lexical field of names and implied identity attributions. On the other hand, names permit strategic moves within the naming field (shall I call him by his first or last name?). They also invite battles over norms (shall our children inherit both parents' names or just the father's?). Names are loaded with precepts and affects, and they can be used to express feelings and judgments towards others. "Name-calling" is one possible result of feelings and judgment intervening in the order of speech.

A good example of name-calling that comes to mind consists of animal metaphors commonly used in Canada. As some readers may know, the frog motif is widely used in reference to French speaking residents of the country. Given this zoological codification of social differences, the question is what animal symbols stand for other segments of Canadian society? Part of the answer to this question lies in a small book, a kind of Canadian version of *Animal Farm* bearing the title *Frog Fables and Beaver Tales* published in the early 1970s by newsman Stanley Burke and cartoonist Roy Peterson (1973). This book and its sequels (1974, 1981) tell the history of relations between francophones and anglophones in Canada, using frogs and beavers to symbolize French and English Canada, respectively. The stories, which have sold widely since their introduction into Canada's political scene, narrate various troubles in a swamp that happens to be called Canada.

Why frogs and beavers to stand for the two official language groups? The French-frog connection is found elsewhere in North America and Europe and is thought to originate from seventeenth-century British allusions to the frog-eating habits of the French. Although less often applied to English Canada,

Cover of *Frog Fables* and *Beaver Tales* by Roy Peterson
and Stanley Burke.

the beaver motif harks back to a fur trade economy and related
imageries and memories of Canada's colonial beginnings. A
beaver on a dam thus appears on one side of the Canadian
nickel, providing a diacritic "coinage" of Canadian identity.
Commercial ads currently shown on television (selling Molson

Canadian beer) also establish explicit ties between Canadian identity and the beaver.

This animal symbolism *à la* Canadienne features four signs consisting of two differential pairs merged into a syncretic correlation: frogs are to beavers what French is to English Canada. To these explicit sign actions should be added all the meanings "automatically" evoked by the frog-beaver imagery. Implicit differences can be organized into the following table of correlations:

Code	Group 1	Group 2
linguistic	English Canada	French Canada
animal	beaver	frog
species	mammal, walks, hand-like forepaws	amphibian, jumps
habitat	woodlands, clear-water lakes, rivers, streams; wooden lodge	shallow ponds, marshes; without shelter
skin	thick fur	spotted skin, slimy
value	trapped, valuable (staple fur, castoreum)	useless, food considered in bad taste
behaviour	skilled builder of dams, lodges, canals; looks after young; silent, modest	fly-eater, idle, toady parasite; does not look after young; noisy, croaky
colloquial	hard-working, eager and conscientious	self-indulgent, scheming (*grenouillage*) contemptuous, bragging, toad-like (*crapaud*)

Apart from being animals that dwell in nature (Canada's wilderness *oblige*) and that live both in water and on land, beavers and frogs have little in common. Their differences attract more attention. As with humans, beavers are mammals that walk and have forepaws they can use like hands, whereas frogs are amphibians that jump. Beavers reside in wooden lodges of their own making and inhabit woodlands and clear-water lakes, rivers, and streams. By contrast, frogs do not build shelters and are found in shallow, murky ponds and marshes. Beavers look after their young; they are the source of valuable

A frog named René (Lévesque). Cartoon by Roy Peterson, Burke and Peterson, *Frog Fables and Beaver Tales*, 16

products such as fur, medicine and perfume (castoreum); and they are reputed as skilful builders of dams, lodges, and canals. Frogs have all the opposite attributes. They do not look after their young. They have slimy skin and are of no value as food, save perhaps for their tiny hindlegs. They are perceived to be noisy creatures idly sitting on lily pads, croaking and catching insects with their protruding tongues.

Colloquial expressions associated with the two animals reflect these differences. An honest, conscientious, and industrious person who works and behaves like an "eager beaver" differs markedly from someone deemed to be froggy or froglike, hence toady. In French, the word *grenouillage* ("frogging") evokes someone who is without discipline, a person who indulges in hanky-panky and political intrigue and swells up with pride (as in one of Lafontaine's celebrated *Fables*). French and English usages of toad imagery point in turn to a person worthy of contempt, a brat, an obsequious parasite, or a servile flatterer, as in a *toad-eater*. Apparently the latter term evokes a quack doctor's assistant, the one who ate or pretended to eat poisonous toads in order that the employer might demonstrate his ability to expel poison.

Given the right text and context, some or all of the differences listed above can be made explicit, hence "depolarized" through narrative explication. This is precisely what the swamp fables written by Burke and Peterson do. The first fable in particular brings out the respective attributes of the two totemic species. Oppositional attributes are assigned to the Québécois and to the English speakers of central Canada and the federal government leaders that have issued from their ranks. Frogs living in the shallow end of the swamp are pictured as lesser creatures, perfidious, unreasonable, fanatical, and emotional. They indulge in song and dance and behave like parasites, deriving benefits from dams built by the beavers. They are politically croaky and noisy, constantly shouting "sep-ar-ate!" In keeping with this imagery, they are obsessed with issues of vocalization, insisting that others learn frog language and speak it well. They are boastful and take pride in their lily flag, rejecting the pan-Canadian maple-leaf flag proposed by the beavers. Beavers act quite differently. They are portrayed as modest, worthy of respect, peace loving and hard working (cutting trees, building dams). They enjoy prosperity and live in comfortable lodges and wear expensive fur coats (Burke and Peterson 1973).

The frog-beaver tale shows how coherent and far reaching the evocations of a simple imagery can be. It also points to the relative complexity of stereotypes. *Frog Fables and Beaver Tales* draws lines of divergence between the industrious and the lazy, the valuable and the useful, the modest and the boastful, the English and the French. This diacritic language is synthesized into an act of condensation that packs many attributions into simple identities. The work of condensation is all the more feasible as the script plays on familiar associations, "depolarizing" connections already known to the audience, such as between beavers and hard work, or frogs and idleness. But the script also appeals to the readers' sense of "creative imagining." Through allegory, it creates new "signaptic" connections and constructs novel sign pathways. When mapped onto events of Canadian political history, stories and cartoon illustrations of beavers and frogs in a swamp do require imagination – the power to syncretize sign actions otherwise mapped onto different planes (polity versus zoology).

But there is more to these frog-beaver fables than "cognitive parallels" drawn between differences in the animal kingdom and divisions in social history. There are also vertical forces at play. The symbolism examined above is rife with normative and emotive associations designed to provoke reaction. Predictable responses to these provocative imageries range from amusement and laughter to judgments in which readers condone or condemn the tales themselves or the conduct attributed to the animals and the groups they represent. The stories are so partial and biased that readers cannot be indifferent to the text, especially if they happen to be actors themselves, readers with inside knowledge of the "troubles plaguing the Canadian swamp." Although all characters are the object of ridicule, little analysis is needed to detect the tendency for some signs to be ranked above others; preferential treatment is given to anglophones and related beaver symbolism, as opposed to the derogatory imagery of the French frogs. The books were clearly written from an anglophone perspective and reflect the corresponding angle on events of the Canadian scene of the 1960s and '70s.

One way to respond to these fables is to argue that social conflicts translated into animal codes betray human thought reduced to its simplest and most objectionable expression: stereotypes and prejudice. But the narrow-mindedness typically attributed to this form of language is no excuse for oversimplifying the sign activity by which oversimplifications are arrived at in the first place. Critiques of mis-representational bigotry should be careful not to underestimate the density of evocations that cartoons can trigger, contentious or not. They must take care not to minimize the cognitive, normative, and emotive processing required to produce caricatures of social life. Nor should they exaggerate the rigidity of conventional thoughts cast from familiar moulds. More often than not, symbols lend themselves to assemblages that are surprisingly malleable and that are subject to rebuttals involving counter-usages of the same names.

Signs are playful, with the implication that they are politically malleable. Broadly speaking, sign manipulations can go into one of two political directions: a mediation and attenuation of opposites (a syncretic tactic), or a full inversion and subversion of the ruling order (a diacritic tactic). Images of frogs deployed in *Frog Fables and Beaver Tales* fit into the broader politics of sign mediation. As with most fables, Burke and Peterson's story of the Canadian swamp is an exercise in mediation that goes beyond static social dualities projected on to the animal domain. Beavers and frogs are opposed in all possible respects. But the story would exercise little appeal were it not for the appearance of a mediator making his entry midway between the beginning and the end of the tale. The character in question is called Peter Waterhole (Pierre Trou-d'eau!), a "wonderful Frog" portrayed as the rich, handsome, well-travelled and knowledgeable "Chief Minister of all the swamp." The lady beavers refer to him as their "Prince Charming," a frog-to-prince conversion echoing Grimm's tale of the frog king (a story of love restoring a frog to his former condition, that of a prince). Thanks to Peter Waterhole, the frogs' ability to jump is turned into an explicit show of agility and prowess. Moreover, unlike all other frogs in the fable, Peter Waterhole is portrayed like the beavers standing up and walking on two legs. Also, he is well intentioned and full of bright ideas. Peter Waterhole thus brings together attributes of the two social-animal species occupying the front stage of the fable. He offers hope and promise of reconciliation in a world

Peter E. Waterhole (Pierre Elliott Trudeau). Cartoon by Roy Peterson,
Burke and Peterson, *Frog Fables and Beaver Tales*, 26

otherwise divided by language and culture. Other signs of mediation include beavers learning the frog language (somewhat
unsuccessfully!), and the frogs learning how to build a dam by
themselves.

Lines of divergence can be downplayed through mediation
and attenuation. But signs of a divided world can also be coun-
tered by means of inversion and subversion. Even when assigned
to French Canada, the frog motif can serve radically different
functions, provided the right adjustments and transformations
are made. An example of this can be found in the well-known
"Frog Song," written by Robert Charlebois in the heyday of
Québec's nationalist movement. In a telling refrain, Charlebois
sings "You're a frog, I'm a frog, kiss me, and I'll turn into a
prince suddenly. *Donne-moé des peanuts j'm'en va te chanter
'Alouettes' sans une fausse note.*" The song goes on to describe
a mediocre, working-class, poorly fed Québécois exploited by
a wealthy boss who makes him work for peanuts and puts him
out of work. The "Frog Song" ends with satirical evocations of
the poor man's good manners and disinclination to blubber and
complain (*"j'te trouve ben élevé pis tu chiales pas"*). Given some
peanuts, he can be counted on to keep on singing his cheerful
"Alouettes."

The lyrics, slang and accent (e.g., *toé* instead of *toi*) displayed
in this song capture key elements of Charlebois's contribution
to Québec's "Quiet Revolution" of the 1960s and '70s. They
express a popular singer's condemnation of social inequalities
and working-class attitudes of subservience and polite resigna-
tion. They also make a mockery of the folk ways of traditional
French Canada (*Alouettes*) and of an elite's conservative aspirations to the refinement of culture, language, and the arts
imported from France – *le bon parler français dans la belle
province catholique.* In "Frog Song," Charlebois pokes fun at
stories of frogs that turn into princes provided that they are
loved for what they are, hence for their quaint habits, their
colourful folk songs, and their zoo-animal inclination to work
and entertain in the hope of getting a handful of peanuts. Thus
"frog" culture is held up to ridicule.

Burke and Peterson associate frogs with laziness, rebellion, croakiness, and noisiness. Charlebois ties the frog motif to an entirely different set of associations: hard work, subservience, stoicism, and joyful music. While pivotal signs may remain constant, their assemblage can vary and generate slants that go from one pole to the other. Beavers "exploiting" a dam in one story become "exploiters" of frogs in another. One frog imagery subverts another. Each composition keeps the other in check, countering objectionable connections with preferred associations. Sign routes explored in language work at countering offensive pathways going in opposite directions.

Critiques of stereotyped representations of social identities (based on gender, ethnicity, religion, occupation, sexual lifestyle, geopolitical affiliation, etc.) are not without relevance. All the same, they often underestimate the malleability of symbolic utterances and related opportunities to manipulate sign connections. They ignore strategies for reorganizing and redirecting sign connections in such ways as to rebut whatever needs rebuttal and pursue the battles of language on its own territory. Stereotypes loaded with value judgments and negative sentiments are not so rigid that only the educated can deconstruct and demystify them, using subtler, "less symbolic" representations of social reality. Paradoxically, abstract thought aimed at debunking stereotyped imagery tends to foster stereotyped views of the symbolic imagination.

Charlebois's song reconfigurates frog symbolism from a Québécois nationalist perspective. The same reasoning can be applied to popular cultural practices that speak to French Canada's relationship to food and forest. In *Les québécois*, Marcel Rioux (1974) makes passing reference to landscaping and eating habits as ways of defining and contesting a people's relationship to nature and culture. The issue of culture's rapport with nature is relevant to our discussion of the beaver-frog symbolism, which happens to evoke the French fondness for eating frogs' legs and the beavers' wood-cutting and lodge-building activities. Briefly, Rioux remarks that unlike anglophone countryside dwellings, which tend to be scattered and surrounded

with trees preserved or planted, rural francophone houses agglom-
erate near the main road and occupy land stripped of all vege-
tation. Trees are cut down to signify human mastery over nature,
a people's hard-won victory against the harsh conditions of rural
life, or at least hopes to that effect (Rioux 1974: 73, 80). In the
author's words, "although when compared to other people he
is still relatively close to his countryside by his manners, his
language and his values, he wants to remain at a distance from
it [the countryside]; he has not taken enough distance to feel
nostalgia ... he has not gotten to the point of developing a
certain kind of romanticism of nature" (ibid.: 82, my transla-
tion). Landscaping *à la québécoise* thus points to a long history
of farmers, lumberers, and *coureurs des bois* whose lives have
been harnessed to nature, "water carriers and hewers of wood"
closely dependent on their immediate environment and the harsh
conditions it imposes. Given this history, the Québécois show
no propensity to idealize Mother Nature, let alone the earth
they are condemned to work by the sweat of their brow.

Paradoxically, the image the Québécois have of themselves
is much closer to the woodcutting beaver than it is to the idle
frog. Rioux finds another confirmation of this beaver-like will
to overcome nature in the Québécois's compulsive fondness for
manufactured objects. The author observes a propensity on the
part of the Québécois to buy artificial, bright-coloured fabric
(nylon and polyester in lieu of linen, wool, and cotton) and
cheap plastic objects of all sorts (trinkets, swans, flowers) to
decorate the spaces they occupy (car, house, yard). These articles give the impression or illusion of wealth – no longer having
to manufacture objects with one's own hands. The same objects
are less often purchased, worn, or displayed by anglophones,
whose tastes in these matters tend to show more sobriety.

According to Rioux, the French Québécois approach to the
environment rather than reflecting a perception of their factual
relationship to nature instead expresses the will to distance
themselves from the ruggedness and drudgeries of life in nature.
Wishful disconnections from the rawness and rigours of fate
and climate are also conveyed through eating habits. This

brings us back to the issue of frogs' legs, which may have been a delicacy in French cuisine (when served *au gratin*, *à la lyon-naise*, *à la meunière*, or *à la provençale*), but are not a popular dish in French Canada. Culinary habits prevailing among "the frogs" involve a fondness for fine cuisine (developed over the last forty years or so) but also a marked predilection for junk food, which accounts for the derogatory expression "Pepsi and Jos. Louis [tart]" often heard in reference to French Canadians (not to mention the current popularity of "poutine," consisting of French fries covered with cheese curds and deep-brown gravy). When compared to English Canada, the Québécois are renowned consumers of processed foodstuffs like white bread and sugar, or roast of pork and tourtière (meat pie) dishes that require meat well done. In keeping with Rioux's argument, French Canada displays a fondness for food that is so trans-formed as to contain little trace of its natural origins. Nourish-ment fit for humans must leave out the colours and fluids of food in its raw form.

To sum up, Burke and Peterson's tale links social differences to a dual animal classification and related positive and negative attributions, commendable virtues and objectionable vices. The underlying grid is cognitive (with historical references), norma-tive (blaming the frogs), and emotive (comical, satirical) all at once. Two columns of similarities and differences are mapped on to divisions between good and bad. Given its judgmental bias and use of cartoon-like characters, the tale is polemical and invites criticism. When examined more closely, however, symbols deployed in this animal swamp allegory offer lessons of mediation and mutual accommodation: one frog turns into an (ingenious, beaver-like) prince who does everything to save the swamp from succumbing to its internal divisions.

Mediation, however, can give way to subversion. Imageries of frogs and related statements about food and forest are not so rigid as to preclude claims and counterclaims of the moral order. Actually, signs of the good life (expressed through animal symbolism) are so malleable that they can follow divergent routes. With barely a few words, Charlebois's "Frog Song" turns the polemical associations and teachings of the *Frog Fables and Beaver Tales* in the opposite direction, using cognate symbols to ridicule and denounce Québécois subservience to nature and to those who treat them as frogs. Patterns of eating (processed food) and landscaping (felling trees) observed in Quebec of the 1960s also show how signs can generate counterclaiming definitions of a people's attitude towards nature and related aspirations to the good life.

As with the "*no importa*" incident, name-calling practices (frog symbolism) illustrate how explicit signs combine with implicit and illicit connections to generate the weavings and battles of sign activity. Semiosis is a political economy of attentionality. All manipulations of language involve dynamic mixtures of sign connections that are fully attended; implications channelled through automatic pathways; linkages left out because without contextual relevance; and associations actively inhibited due to their offensive character. Sign action circuits are like main roads surrounded with secondary routes and forbidden pathways. Multiple lines run parallel to explicit chains of thought and are blocked off to varying degrees and for various reasons – reasons that may be individual or social, normative or emotive, temporary or permanent.

Heteroculturalism

Sign reticles are horizontal assemblages of similarities (RH) and differences (LH). These assemblages are not static and they are not strictly cognitive. Instead they are highly malleable and invite constant interventions of vertical effects of the normative (prefrontal) and emotive (limbic) kind. The axial intervention of judgment and sentiment implies in turn a political economy of attentionality. Sign politics generate constant efforts to silence or downgrade things that are deemed offensive to morality or the pleasure principle. By the same token, sign activity grants two privileges to indices of "goodness": greater visibility, and the powers of strategic silences – brevity, timeliness, and diplomacy.

The rank ordering of sign attentions affects all compositions of similarities and differences. It also determines the actual weight that needs to be accorded *to either similarities or differences, towards strategic moves of the syncretic or diacritic kind.* Illustrations of syncretic tactics include christening and family naming practices, assemblages that emphasize signs of unity and selfsame identity but that nonetheless insert division and rank order into personal identities (the Christian name comes first, whereas the legal-biological "sur-name" comes last). The hemlock motif of line 1 of *Evangeline* is also part of a composition that elevates signs of unity in time and space (e.g., the evocation of tall evergreen trees) above notes of subsidiary disharmony (e.g., the poisonous hemlock weed). Similar comments apply to our discussion of higher-order terms such as "man," "day," and "life cycle": these "upper-class" terms wishfully unite opposite terms through a preferential treatment that seems to be part of the order of things. "Man" unites man and woman; "day" unites daytime and nighttime; the "cycle of life" unites life and death. The imagery of *Frog Fables and Beaver Tales* also heads in a syncretic direction, towards an overt reconciliation, thanks to the mediation of Peter Waterhole, of divisions and dualities pitting the French against the English. All these are examples of sign tactics paying special attention to lines of unity and mediation. They all do so at the expense of effects of *diakritikos* otherwise required and firmly embedded in the same imagery. Syncretic tactics signify and downplay divisions while discreetly carrying biases of their own.

Illustrations of diacritic tactics have also been explored, starting with the fig apron motif of biblical origins (cf. *3-D Mind 1*). The fig imagery of Genesis 3 was shown to underscore signs of the Great Divide: creator separated from his creation, man from woman, life in Eden from life after the Fall. Themes that bring these opposites together are backgrounded and kept active at the same time. Unifying themes pushed into the background include the notion of creation (making something out of nothing and for no necessary reason) and the idea that painful labour (male work, female travail), be it waived or imposed, is what the creation of man and woman is all about. Shadows of syncretic mediation also include the prospects of redemption and afterlife, themes that are part of the broader story but cannot be attended in scenes that focus on the Fall. (Similar comments apply to the cornchild-iguana battle as told by the Mexican Nahuas. The argument made in *3-D Mind 1* is that hero and foe have much in common despite appearances to the contrary. They share a common aspiration to mediate oppositions in nature seen in a Nahua perspective, those that separate wetness and dryness and also water, earth, and sky).

Other cases of diacritically oriented attentions include the "exploitation" debate, the *"no importa"* anecdote, and the rule against walking under a ladder. The debate regarding what should be understood by "exploitation" involves a radical opposition between two political philosophies, with no apparent potential for a syncretic approach developed somewhere in the middle. Given this overt opposition, views shared on both sides of the divide are bound to pass unnoticed; a shared commitment to rules of "methodical mastery" ("man" by "man" in Marxism, "nature" by "man" in bourgeois economics) is thus silenced and becomes a non-issue. The "*no importa*" incident is diacritically skewed as well. It negates the similarity that may lie between worries at home and greater sufferings endured by others. The negation, however, is of a wishful nature; it betrays an on-going struggle – difficulties in separating things that matter from those that don't. Lastly, the curse that comes from walking under a ladder separates signs of good luck reached "at the top" from signs of bad luck experienced at the bottom and underneath. The antinomy proposed here takes the ethics of luck and fortune (and its spatial representations) as a given, an overall perspective on life not meant to be explicated let alone demonstrated or questioned (could it be that greater rewards may be obtained by accepting lowly positions, as in the ethics of sacrifice?). Attention is given to the opposition between good luck and bad luck, but not to what these two things have in common – the teachings of "fortune."

Sign actions fit into attentional strategies heading in syncretic or diacritic directions. Individual signs are nonetheless malleable in that they can serve different masters, depending on the circumstances. Given the right rearrangements, signs committed to particular interests can be subverted and subsumed under a scheme that serves opposite ends. Our discussion of Charlebois's frog song thus showed how signs of popular culture, stereotyped and cartoonish as they may be, are never wedded to one strategy alone. *Synkretismos à la* Burke and Peterson features a beaverlike frog bringing harmony to the swamp. Through parody, Charlebois's frog imagery fosters instead a nationalistic reassessment of marks of difference and distinction, away from sentiments of unity achieved through political subservience and folkloric docility.

One general conclusion that follows from these analyses is that signs of name-calling (e.g., frogs) have much in common with acts of naming (e.g., Jacques Chevalier) or calling (e.g.,

the word "man" used to call both man and woman). They all
lend themselves to expressions of feelings and judgments that
assign attention wherever attention is most wanted, be it in the
direction of greater distance and difference (diacritic tactics) or
closeness and sameness (syncretic tactics). The end result is an
attentional polity that affects all moments of semiosis, including
those that may seem neutral (using the word "day" to stand
for the unity of daytime and nighttime), trivial (a personal
anecdote, a superstition concerning ladders), fictive (the corn
myth), cryptic (the fig apron), or polemical but nonetheless
amusing (frogland mythology).

Debates over the meaning of what we choose to call "man"
or "exploitation" remind us that sign tactics will also feed into
situations of "real life" politics, including regimes of domination and acts of resistance. An offshoot of naming and calling
practices, *identity rhetoric* is a particularly good example of
sign activity that feeds into events of political history. In *3-D
Mind 1* I suggested that the power of identity politics (based
on nationality, ethnicity, gender, sexuality, etc.) lies in its capacity to reinforce *or* destabilize existing regimes, not in its truth
value understood in a representational perspective. I added that
when contributing to progressive movements, identity struggles
may use the language of "strategic essentialism" to draw attention to fictions of stable homogeneity, a syncretic language
adapted to the promotion of multiplicities otherwise repressed.
I now suggest that the opposite scenario is equally plausible.
Twentieth century history is crammed with examples of constructions of identity serving dominant interests and the pursuit
of power based on the rule of oppression.

This brings us to the use and abuse of the rhetoric of difference – a diacritic tactic – to account for inequalities and armed
conflict in Africa, with particular reference to official accounts
of recent warfare between the Nuba and the Baggara in Sudan.
Briefly, wars in Africa are often explained away as ethnic conflicts, wars of religion and tribal identity. As Suliman (1999)
argues, the problem is not so much that the explanation is false,
which it is. The problem rather is that the explanation has a

tendency to make things worse, fueling the conflicts, and not without intent.

Up until the 1980s the Nuba and the Baggara were relatively at peace. Since then they have been at war with one another. The civil war that broke out in 1983 led the Arab Jellaba government and eventually the National Islamic Front to repress the Nuba-led opposition party called the Sudan National Party. They also armed the Murahaliin militia and the Baggara nomads against Nuba communities and the Sudanese Popular Liberation Army roaming in the rebel-friendly Nuba mountains. Faced with problems of overgrazing and persistent droughts, the Baggara used this opportunity to raid Nuba communities and dispossess them of their land. These raids, however, have benefited mostly the Jellaba government and a minority of land-hungry Jellaba farmers and absentee landlords intent on introducing large-scale mechanized farming into the sub-humid Nuba mountains, with financial support from the World Bank.

Not surprisingly, the official account of the Nuba-Baggara war is quite different and revolves around issues of "difference." While they actively supported the Baggara war against the Nuba, landlords and the government fueled the conflict by treating it as both an outburst of tribalism and a jihad or holy war against the non-Islamic Nuba. War boils down to the fact that people are different and hold on to their differences, as opposed to all the things they differ on or about.

Signs of identity and difference that are used to account for ethnic populations at war assume three lines of reasoning. First, a group is delimited by recognizable boundaries. Second, its identity is constituted by what is shared between members located inside those boundaries. Third, differences result from comparisons drawn between group identities. Attention granted to marks of difference and related conflicts implies that the "relational aspects" of broader social systems can be ignored. Silence on these larger issues becomes the rule. No importance is given to the interactive dynamics that constitute the "inside" in two ways: through exchanges among those deemed to be alike, and through interchanges with the "outside" world. The implications of this twofold silence should not be underestimated. Viewed from a "relational" perspective, life in society can be argued to be neither monocultural nor multicultural. Each society does not simply have a culture of "its own," a collective identity produced by itself and for itself, to be exported and communicated to "others" when and if need be. Nor can we reduce societies to variable sums of multi-ethnic and pluralistic island-like "selves." The relational foundations of social life imply rather what might be called a heterocultural or heterosocial rule. Stated simply, this principle means that social life thrives on "intercourse" (a syncretic term *par excellence*) amongst those deemed to be alike (all the Nuba people) and also between those deemed to be different (the Nuba and the non-Nuba).

What does heteroculturalism look like in real life? Consider the Sudanese Nuba. They are Nuba for several reasons. First, they are what they are because of the particular *ways in which they construct differences and relations*. They demarcate themselves through methods of establishing divisions and links between villages and lineages, the young and the old, men and women, people and land, plants and animals, humans and spirits. Second, they are Nuba because they know *who they are not*: namely, the neighbouring Baggara who construct differences and relations differently. Nubaness signifies how the Nuba do things in a non-Baggara (or Jellaba) way. Third, the Nuba are Nuba not because they don't mingle with the Baggara. Rather their identity is a product of the intercourse that binds and sustains the two populations: real commerce, intermarriages, and so on. Paradoxically, at the heart of Nuba identity lies a long history of trade and politics linking the Nuba and the Baggara. Interdependence across this ethnic divide includes nineteenth-century stories of Baggara subtribes defending their respective Nuba hills and allies in order to secure supplies of grains and slaves.

How does this discussion of Nubaness illustrate the idea of heteroculturalism (and its tactical negation)? To begin with, the analysis suggests that Nuba identity resides in a web of negotiated differences and relations, internal and external. *Exchanges within and between* the two bodies, the Nuba and the Baggara, determine the shape and the anatomy of each group. Another important point to be made is that "relational identities" are in constant motion, something that the concept of fixed differences tends to hide. The Nuba are Nuba not because of static attributes that can be assigned to them but rather because of convoluted stories of relations changing over time.

Heterocultural identity formation is grounded in the internal and external reciprocities of social life. I hasten to add that "attentions" of the body social are not made up of "thoughtful considerations" and "mindful reciprocities" alone. Social identities are also covered with wounds suffered at the hands of other groups. Less metaphorically, the Nuba history of ties with other groups includes stories of slavery and repression verging on genocide. "Relations" binding this group to its "significant others" have ranged from commerce and marriage to invasion, armed conflict, oppression, and the impositions of social and economic inequities of every sort. All fetters and bonds, be they cooperative or conflictual, play a direct role in stories of shifting identities.

Without outside interaction, Nubaness may not have been constructed as a distinct cultural identity. The Nuba comprise more than fifty dialect groups who share above all a common history of Turkish and British invasions, Jellaba domination, and slave raids at the hands of the Baggara. Previously roaming the plains of Kordofan, the Nuba were forced by continuous raids to retreat into the Nuba mountains; territorial identity was not and is still not a simple matter of each group choosing its habitat independently of outside forces. Without this common destiny vis-à-vis external forces, the boundaries of Nuba identity would be meaningless. Even the term "Nuba" has been imposed from outside (as is often the case). Ethnicity is never merely an internal construction. It is also a response to external actions and definitions.

Recognizing interdependence is all the more important as it can serve the cause of peace. The Nuba and Baggara undertook peace negotiations in the early 1990s. Both populations came to the conclusion that they stood little to gain from the armed conflict and preferred instead to resume trade activities: Nuba-grown cereals in exchange for salt, clothes, and industrial goods imported from Kharfoum by the Baggara. But the market economy has also its downside. Nuba identity has been affected by recent population movements, urban migration, and increased trade with other groups. Do these trends towards further outside interaction contradict our "heterocultural" thesis? Not necessarily. Urbanization is affecting the Nuba way of life but mostly in the sense of making constructions of ethnicity more rigid than ever. According to Suliman, the Nuba have further "discovered" their Nubaness through the diaspora. Life in towns of the Sudan and expressions of northern Arab arrogance towards non-Arab southerners and westerners have reduced Nuba cultural diversity to a single second-class Nuba identity. Through official discourse, people are pigeonholed into ethnic categories, with the result that "culture" is artificially disembedded from broader forces and networks of social history.

Students of ethnic identities in Africa must be careful not to play into the rhetoric of "tribal difference." They should not ignore the relational politics that shape group identities and bind them to one another, for better or worse. Sadly enough, the history of social sciences has in common with colonial and neo-colonial politics a propensity to divide or reorganize populations into apparently homogeneous national, ethnic, or linguistic groups. People are slotted into island-like entities that hide the interchanges and movements occurring across boundaries. "Communities" are frozen into biological-like organisms, to be classified like genuses and species, divisive categories feeding into strategies of domination and war. The end result is a hierarchical and conflictual brand of heteroculturalism hiding under a thick cloak of "tribalism" – people fighting apparently because they hold on to their differences.

This ends our brief illustration of an African embattlement of sign activity. We now proceed to other illustrations of the rank ordering of signs, this time with an emphasis on the battles of morality and the attentions of Eros.

The High Road and the Low Road

Sign connections and attentions are malleable, lending themselves to syncretic and diacritic tactics and the struggles that follow. These tactics skew the ordering of similarities and differences, using the attentions of language to achieve affective resonance and moral import. The outcome is a rank ordering of signs based on regimes of attentional merit, a logic of stratification that plays on three levels: the explicit, the implicit, and the illicit.

noticed	explicit	expression	depolarization
licit	immediate	exhibition	somatic
unnoticed	implicit	impression	depolarization
licit	proximal		automatic
unnoticed	illicit	repression	hyperpolarization
illicit	distal	inhibition	

1. Explicit circuits permit attentions to be focused on pivotal meanings (cognitive), affects (emotive), and precepts (normative). They constitute the main-track impulses of sign activity, "somatic" connections depolarized and exhibited for purposes of "signaptic" communication.

2. Implicit connections are depolarizations attained through automatic pathways, sign impulses carrying lower measures of attentionality. They constitute secondary pathways that hook up things implied to things explicated. Their role is to support main-track effects with proximal meanings,

sentiments, and morals that need not be overtly expressed
and that may be left unsaid or unnoticed.

3. But things left unnoticed also consist of illicit associations,
off-track and menacing pathways that must go unnoticed
and be blocked off through higher levels of "capacitance" –
sustained efforts of repression, inhibition, or hyperpolarization.

This is to say that sign linkages are unequally attended, rank
ordered according to the relationship they entertain with acts
of attentional noticing. Language is a value-oriented logic that
selects a limited set of connections for somatic depolarization,
producing immediate sign actions that expel all other same-field
connections from the surface narrative. Proximal associations
are processed through graded implications channelled through
automatic pathways. As for distal linkages, they are bypassed
or distorted through circumvolution, via the inhibitions of language. Through efforts of hyperpolarization, active tensions
and anxieties are both reflected and deflected in semiosis.

The illustrative analyses presented below expand upon this
threefold approach to sign activity. But they also challenge some
commonly held views regarding the relationship between acts
of attentionality and the teachings of moral consciousness. I am
alluding to the widespread idea that signs concerned with right
and wrong are necessarily attended by consciousness and made
explicit in language. This means that the principal activity of
consciousness lies in expressions of morality. By contrast, the
unconscious is considered to be emotive and immoral, the realm
of illicit stuff that must be left unnoticed. What we end up with
is another kind of half-brain talk, this time mapped on to the
vertical plane. The whole brain is so organized that the lower
limbic system does things that are fundamentally different from
what prefrontal lobes do. Both may be in the business of pursuing things that are deemed to be desirable, yet desirability is
fundamentally a *diacritic* affair: the sign divides into two branches
as soon as it is uttered. A bifurcation written in the wax of the
brain requires signs of goodness to travel either on the high
road of conscious morality or the lower road of limbic desire.

Convential wisdom portrays the desirable understood morally
as the mirror image of the desirable attained without permission.

Things that are wished for and are morally unacceptable (e.g., sexual pleasures) are pitted against things that are morally prescribed and limbically unattractive (e.g., sexual abnegation). These binaries are hard-wired into our brain, which means that subjects are obliged to attend either the ethical or the pleasurable – either the rulings of law or the powers of desire. As we all know, the dilemma contains a foregone conclusion: "thank goodness," normal subjects will *attend* the moral side of things and give them proper *expression*, at least to the best of their human ability. The upshot of this reasoning is that morality becomes inherently inhibitive, never to be inhibited. By constitution, morality is channelled through acts of explicit consciousness and has no constitutional reason to hide.

Moral sign activity, however, is not reducible to declarative and repressive pronouncements of the law. The interfacing of precept and affect does not lend itself to dualistic models that simply pit overt morals against covert pleasures, or the attentions of law against the devious motions of desire.

The idea that conscious morals (of prefrontal origin) are constantly battling with unconscious "impulses" (of limbic origin) can be challenged on several grounds. One reason has to do with the double-bind effect of moral utterances that are compelled to speak up against the unutterable. The point has already been made. Briefly, we have seen how the anxieties of sign activity will stir language in morally acceptable directions, reducing the motivation to proceed along forbidden chains of thought. But connections that speak of libidinal pleasure (or fear of displeasure) and that verge on immorality (or tragedy) cannot be ignored. Things deemed unutterable must be given some prefrontal attention if they are to be condemned, relinquished, or avoided. Also the unutterable must be explored and kept extant (as when the author of Revelation describes the fornicating ways of the whore of Babylon) for the lessons it offers (retribution on Judgment Day). In the end, some concessions must be made by prefrontal norms to signs of limbic activity.

But there is another factor that pushes morality into making room for pleasures and fears of subcortical origin: the need for morality to enhance its own appeal, adding lure to the rule of

law, as it were. In and of themselves, pronouncements of the
law are unattractive. Rewards and compensations of all sorts
are needed to seduce the mind into accepting duties and obligations of moral conduct. This is to say that moral virtue will
receive attention provided it appropriates some of the "virtues"
of desire, the pleasure principle, and the avoidance of pain.
Take the *"no importa"* incident. In hindsight this is a case of
a moral injunction (don't make a mountain out of a molehill)
expressing itself through convoluted imagery (ice cream on my
shirt ... *no importa*!) so that the mind can fight off unpleasurable thoughts (professional worries).

Morality in semiosis is not driven by a simple hermeneutic
function, which is to spell out right from wrong and good from
bad. Signs of moral import must act in connivance with the
powers of affect if they are to attract attention and win over
the "lure of evil" (i.e., pursuing pleasure and avoiding pain at
any cost). But the obverse is also true: in order to find satisfaction, affects must appropriate some signs typically associated
with normative activity, especially those having to do with acts
of *deprivation* and *dispossession*. Paradoxical as it may seem,
*the virtues of power and pleasure must seek the powers and
pleasures of virtue*. If pleasure is to maintain the mind's attention beyond short spaces of time, it must borrow some tools
of normative language, such as deferment and self-denial.

In analyses to follow, we shall see that emotional activity is
not reducible to the undelayed satisfaction of limbic wants.
Rather, affectivity thrives on "painstaking" yearnings that
entail sacrifices of two sorts. First, the cravings of desire presuppose losses incurred in time, through the deferral of pleasure; more shall be said about this issue in *3-D Mind 3*. Second,
losses are also incurred in space, through dispossessions of the
body that spur the longings and appetites of desire. Manifestations of loss in time and space preclude clear-cut distinctions
between depravation and deprivation – sentiments based on the
desire to possess versus norms inspired by acts of dispossession.
These divisions do little justice to the weavings of affect and
precept. Concessions to limbic sign activity are to be expected
from acts of speech guided by normative schemes and codes.

Likewise, signs loaded with limbic affect are bound to make concessions to mechanisms of deferment and renunciation. Signs of emotivity are known to turn wants (feeling hungry) into the pleasures of desire (whetting one's appetite).

The essays that follow explore exchanges between the lawful and the pleasurable, using illustrative material based on constructions of gender and sexuality. Our findings will reinforce some of the points developed in previous analyses concerning the freeplay of sign activity. The emphasis, however, will be on *downward allowances* that acts of judgment make for signs of limbic activity, and *upward allowances* that acts of desire make for signs of deferment and self-denial. We shall see that far from being mutually obstructive, connections between pleasure and morality can be mutually enabling. Traces of the pleasurable (verging on evil) can be found in the pathways of morality, and traces of the ethical (bordering on sacrifice) in circuits of licence and depravity.

Intersections and interchanges between levels of signification will be highlighted, over and beyond a mere classification of functions in semiosis. We know that variable amounts of attention can be granted to either affect or precept, pleasure or suffering, self-gratification or self-denial. But there is more to semiosis than digital either/or choices applied to dichotomies mapped along the normative/libidinal axis. As in neural activity, analogical (ratio-based) combinations of sign functions are the rule, not the exception. Sign actions are constantly intermixing desire and norm, gratification and frustration, acts of indulgence and calls for renunciation. Semiosis combines these functions to varying degrees and with malleable outcomes. This may result in events that are *predominantly* moral or emotive, yet the end product will always contain some admixture of complementary ingredients receiving both attention and inattention – the most elementary events of sign and brain activity. This last point is crucial to our argument. Given their multiple meanings (Rosenzweig et al. 1999: 202), attention and inattention constitute possibly the best examples of what I have called "flexible simples" (see *3-D Mind 1*). That is, they presuppose interactivity. As such, they cannot be confused with fully fledged tasks (moral

	pleasurable		unpleasurable	
	joyful (good)	sinful (evil)	tribulational	sacrificial
possession	option a	option b	option c	option d
dispossession	option a^{-1}	option b^{-1}	option c^{-1}	option d^{-1}

discourse, emotional impulse) and coincide even less with concepts of western inspiration (for instance, the notion that consciousness is attention paid to rational and ethical norms and related constraints on the satisfaction of immediate wants).

Discussions to follow offer interpretive analyses to illustrate the two-way concessions of morality and the pleasure principle, using imageries of footwear and lower limb anatomy found in the Bible and in modern and postmodern scripts as well. Analyses of biblical imageries of loin, hip, knee, foot, shoe, and door are presented with a view to showing how precept comes together with affect (or suffering with pleasure) to produce intricate assemblages of sign activity. Four compositional outcomes will be studied, using themes already introduced in previous analyses: the joyful, the sinful, the tribulational, and the sacrificial. These variable scenarios will illustrate the interplay of two semiotic theorems. In keeping with the *sign malleability theorem*, all four scenarios will be shown to recuperate and accommodate the same foot and shoe imagery. What is more important, however, is that these symbolic manipulations lend support to our *morality-immorality exchange theorem*. As explained below, they illustrate how pleasure (either joyful or sinful) must incorporate signs of dispossession and, conversely, how unpleasure (either tribulational or sacrificial) must be signified through acts of possession.

At first sight, the biblical foot and shoe imagery gives an overall impression of rhizomatic arbitrariness; anything goes, so to speak. On the one hand, imageries of biblical shoe possessions can accommodate the pleasure principle viewed morally or immorally. Thus, when associated with blessings of wealth and property sanctioned by God, the shoe is synonymous with blessings of the *joyful* life (option a). If abused and undeserved, however, the same footwear will evoke *sinful* behaviour, or an excess of pride and property embodied in shoe-like possessions of this world (option b). On the other hand, signs of shoe possessions can be made to move in the opposite direction. They can step into negative or unpleasurable spaces, hardships that once again may point to moral or immoral conduct. Thus, when worn, shoes may trigger fears of *tribulational* events: for instance, the punitive pains of deportation and separation from family, land, and wealth (option c). The same footwear, however, can also connect with the teachings of *sacrificial* virtue and faith in God, to produce the imagery of exemplary preachers wearing shoes in exile (option d).

The footwear motif can follow any of these four pathways. The dominant meaning actually triggered hinges on the details and connections made explicit in a given scene. For instance, the command to "eat like this; with a girdle around your waist, *sandals on your feet,* a staff in your hand; you shall eat it hastily: it is Passover in honour of Yahweh" (Exod. 12.11) clearly involves a ritual expression of readiness for a sacrificial journey prescribed by God. Other pathways potentially triggered by the shoe possession are blocked off from this particular text, to be explored and expanded in other scenarios of biblical inspiration. Exodus 12 chooses one plan, the sacrificial, and follows it through.

Imageries of shoe possessions and their variable associations – the joyous, the licentious, the calamitous, and the virtuous – illustrate the flexibility inherent to semiotic activity. The principle of sign malleability does not stop here, however. Curiously enough, the variegated pathways we are about to explore can be accessed through the inverse imagery: that is, removing the shoe as opposed to wearing it. The same effects – pleasure or unpleasure for moral or immoral ends – can be obtained by ceding the shoe instead of seizing it, letting it go instead of grabbing it. As with feet covered with shoes, the foot can be uncovered for either good reasons (marriage consumed, option a^{-1}) or evil purposes (going to bed with a whore, option b^{-1}). Also it can be withdrawn with either tribulational implications (being deprived of one's shoe property, option c^{-1}) or sacrificial intent (removing the shoe in a sacred site, option d^{-1}). The question as to whether a character possesses or is dispossessed of the shoe certainly matters to the composition at hand. Both answers to the question, however, can be adjusted to any of the four pathways outlined above.

To use the language of neuropsychology, all is as if both strategies of RH withdrawal (ceding, letting go) and LH approach (seizing, grabbing) could equally apply to limbic and prefrontal expressions of things deemed worthy of desire or fear. An approach attitude entailing possession of the shoe can have the same overall effect as a withdrawal attitude implying dispossession. Opposite strategies, those of covering or uncovering the foot, are compatible with signs of moral edification and limbic appeal alike. Far from being concerned with denotation or symbolic codification, semiosis appears to be an utterly playful machine. Postmodernism may be right after all.

Still, these observations beg the question: Why does sign activity need so much flexibility and freeplay? Isn't there a risk that the infinite sliding of signs will foster *l'indifférance*? Also, how do we account for signs of possession to convey morals of sacrificial behaviour and related losses of worldly belongings? Why should a man pursuing God's ministry in exile be allowed to wear shoes? Should he not simply divest himself of all things worth coveting and the symbolic expressions thereof?

The answer to this riddle lies at the heart of brain and sign activity – the dialogue that connects *metaphor* to the order of desire (normative and emotive). Briefly, we know that a metaphor establishes a relationship of similarity between things different in other respects (see *3-D Mind 1*). A metaphor that likens A (Christ) to B (a lamb) expresses first and foremost a relationship of similarity. But the act of metaphor is essentially a syncretic weaponry, a tactic that requires a subsidiary recognition of *diakritikos* – some concession to the distance that lies between things that are similar but never identical (faith in Christ is not animal worship). Through metaphor, A is likened to B so that it can be substituted for it *but cannot take its full place*. Things cannot be said to converge through metaphor if they are in a state of fusion or indifferentiation. To use the biblical footwear example, a shoe can be shown to resemble and stand for wealth, property, land, house, and womb. The shoe is nonetheless different from all other possessions and is generally worth less compared to the objects it stands for.

Given these ambivalent features, the effect of "substitution" achieved through metaphor can take one of two opposite directions. Either the metaphor marks and appears together with the objects of desire it stands for, thereby adding on a symbolic "representation" to what is being "represented"; a man can thus ritually acquire new shoes at the same time as he "takes possession" of wife and real property in land. Or the metaphor may be granted *in lieu of what it stands for, as a sign of what the subject longs for* – shoes as a pale reflection of the real possessions they simulate and that may be lost (because of punishment or sacrifice). The two scenarios revolve around the shoe-worn metaphor but generate radically different compositional effects. The first scenario attracts attention to the syncretic principle that unites the symbol and the possessions it evokes and comes with. The second stresses instead the diacritic qualities of metaphor, simulations written over bodies "possessed" with lack and desire.

Much shall be said about the second scenario and the implications it has for a theory that considers the seductive ways of simulation and metaphor. Suffice it for the moment to stress that by metaphorical simulation is not meant a final possession. Imageries that feign commerce between the desiring and the desired – say, for instance, erotic interchanges between the masculine and the feminine – point to lack and want and related things that are longed for. They convert the sufferings of want into the pleasures of desire.

Enough has been said about what has yet to be explained in detail, through a close reading of foot and lower limb imageries in the scriptures and in present-day expressions of popular culture. Our study of how pleasure links with morality begins with the story of Jacob and his red-haired twin, Esau, a sibling relationship portrayed through the anatomy of heels, hips, and knees.

Jacob the Heel-Catcher

All sign linkages are endowed with the powers of movement and flexibility. The same can be said of articulations of the body. Heels, knees, and loins permit the body of language to take many courses of action. Depending on the corpus at hand, the pathways of sign attentionality may range from communications of joy and sin to lessons of suffering and sacrifice. When steps are taken in one direction (say, the joyful), other trajectories (say, the sinful or the tribulational) are banned from the surface plot. Impulses proceeding along main and secondary tracks can be activated provided that sidetrack potentials are blocked off, if only for a narrative while.

Consider the story of Jacob, conveniently narrated in five "steps."

Step 1: The bright side of Jacob's story can be told via the erotic and joyful associations of the hip, the knee, and the heel.

In the scriptures, the hip is tied to the loins and the lower abdomen, parts of the body regarded as the region of manly vigour and procreative power, and requiring clothing. Hips have in common with knees that they provide *Homo erectus* with a sturdy back and the solid legs and thighs of a powerful begetter. Accordingly, the grandchildren issued from a man's loins are proverbially placed on his lap or knees, which is where the strength or the weakness of a man lies. Israel's founder was blessed with prolific loins in that he was an exceptional begetter.

God said to Jacob: "Be fruitful and multiply; a nation and a company of nations shall be of thee, and kings shall come out of thy loins" (Gen. 35.11). Note that men with sturdy loins and knees are likely to be "well-heeled" in the sense of possessing wealth and being armed. In a now-archaic English usage, to heel someone was to equip or supply a person with something, especially money.

Step 2: Although enviable, the powers vested in loins, knees, and heels can lead to sinful behaviour.

These are powers that cannot be enjoyed without risk. Given too much wealth and military strength, humans can fall prey to temptations of pride and rebellion, lifting their heels against rightful rulers or God (Ps. 41.9, John 13.18). A human body erected above God's creation points to an act of treason. The crime is reminiscent of the betrayal of the Almighty Creator by the demonic prototype of someone acting as a "creep" or a "heel." Adamite successors of Lucifer behave like horses lifting their heels against the Master. They behave like haughty men living for kicks, men heeled with plentiful wealth and the organs of lust, malice, and war.

The wealth embodied in the heel is all the more reproachable as it is an object of envy and enmity and may be acquired through conquest or outright theft. This brings us to the story of Jacob fighting with his twin brother (Gen. 25.26, Hos. 12.4). Esau, also called Edom, was born first and came out of his mother's womb with a red-haired body. Esau means hairy, Edom means red. He was the firstborn heir to his father's blessings, a son with a wild and restless character, a cunning hunter to boot! Jacob was born soon after, with his hand holding on to his brother's heel. He was a man of domestic habits, an agriculturalist who dwelt in tents. The name Jacob means "heel catcher," a character destined to bring his elder twin brother to heel, so to speak. But the name also denotes a person who cheats, defrauds, or deceives. The allusion befits a person who usurped his brother's birthright through imposture and a raw deal (Esau's right of primogeniture in exchange for red pottage). Jacob obtained the blessings of Isaac by impersonating his hairy brother and serving his blind father a savoury dish of kid meat dressed as venison.

Step 3: Tribulation is bound to ensue from powers obtained through sin, without merit.

Esau the hunter treads hard on the heels of the animals he preys upon and makes a living out of killing. Accordingly, the man was born with the expectation of deriving wealth and property from his father's death. As a result, his heel becomes the most vulnerable part of his body, caught from his birth by his own twin brother who ends up usurping his wealth. The heel may be a symbol of power and property, yet the power it represents confers great weakness and vulnerability, especially when possessed without merit or restraint. Trials are bound to follow. Flesh made powerful is as vulnerable as Achilles or Esau. Riches are so futile that everyone should "fear in the days of evil, when the iniquity of my heels shall compass me about" (Ps. 49.5). Wicked men who aspire to become well heeled and follow in the footsteps of the evil serpent are likely to fall where they have sinned. That is, God will tread on their heels and watch every move they make. Their heels will be caught in a snare, the Serpent will bruise the heels of their (wives') offspring, and snakes will bite the heels of the horses they mount with so much pride.[4] Jacob and Dan the serpent can testify to the weakness that lies in worldly strength, in the might of cruel beasts and men trampling their enemies underfoot. Death treads on the heels of the proud and the arrogant.

Although stolen from Esau, the blessings Jacob received from Isaac were sanctioned by God. The man thus became the eponymous founder of Israel, a well-heeled nation he led to the Promised Land. The patriarch, however, suffered many "evil days" (Gen. 47.9). He could hardly seize his brother's heel and property without his own heel being caught in a snare. Jacob started out as a semi-nomad and was later subjected to a twenty-year exile with no permanent land to call his own and to tread with his own feet. He lived a troubled life amongst giant nations multiplying in the land of Canaan.

The troubles faced by Jacob and his descendants find a telling representation in the hip and leg injury he suffered as a result of the battle he fought with an angel. Jacob's loins were marked by pain from the moment he met God via the angel. We know that Jacob prevailed over the angel he wrestled with, yet the struggle left a mark, a permanent limp that warned him of difficult times to come. His encounter with the angel was brought to an end with his hip joint being dislocated by a blow to the socket of the hip. He was hit at the level of the thigh or the sciatic muscle, a piece of flesh of animals that the children of Israel are forbidden to eat. The sciatic nerve passes down the back of the thigh and is attached to the loins and the organs of procreation; it is the most important in the maintenance of lower bodily life. Sciatic pain in the loins results in many cases from a vertebral hernia. More serious injuries to the spinal column can cause paralysis and impotence.

The limp the angel inflicted on his opponent involved a lower-limb paralysis that warned Jacob and his people of pains and troubles on their journey to the Promised Land. Because injured, the patriarch was well prepared for a difficult battle; forewarned is forearmed. Likewise, Adam and Eve's discovery of their shameful nudity is followed by scenes of bodies covered with loincloths and a woman cursed with the affliction of labour pains and travail. Biblical scenes of men placing their hands on their loins like women in labour are associated with ashen faces and visions of future tribulations (Jer. 30.6).

Limping Jacob is not cursed to the point of being stripped of his loins' capacity to multiply and inherit the riches of the earth. Still, the image of a hollow thigh out of joint has serious implications for children of the earth. Man's seat of generative power is here treated as the site of wrenched bones and shrinking tissues that barely unite flesh and bone. It is as if pain alone could unite the solid bones of manly existence to the hollow side of the womanly flesh. The site in question becomes the

weakest spot on the human body, to be girded about for reasons
of purity as well as protection from inevitable sufferings. Weak
men have a soft spot for a bone-like muscle of tender flesh in
the loin region, but the flesh enjoyed for its own sake brings
nothing but sorrow. Human sexuality can never be detached
from the wrenched sinews of travail and war.

The sufferings of life can affect the loins, the leg, the knee,
and the heel. In biblical terms, utter ruin entails pain in the
loins as well as sore knees trembling with fear (Deut. 28.35,
Nah. 2.10). A man forced to fall on his knees before his enemy
will be portrayed with feeble knees or a crippled back. Men
also assume this posture in the presence of the Almighty. The
Father summons men to fall on their hands and knees and to
wait on him hand and foot. Emasculation imposed by God or
foe hits all articulations of the lower body.

Step 4: Given some adjustments, sufferings of the lower body
can be given a sacrificial twist.

Provided they are fully assumed, trials and tribulations afflicting the lower articulations of the bone-and-muscle structure can
be converted into acts of abnegation. Painful disarticulations
are conducive to an ascetic treatment of the human body. Biblically speaking, a man wearing a girdle about his loins may be
carting around wealth in his lap, but he may also be preparing
himself for a difficult journey and showing courage in the face
of adversity, as with Jacob after his encounter with God. The
biblical command to "gird thy loins like a man" (Job 38.3,
40.7) thus refers to the fastening of loose-flowing robes with
girdles or sashes when men ran, laboured, or fought.

Knees also lend themselves to expressions of sacrifice. A
father must bend his knees and lose part of his strength if
children are to sit on his lap and attest to his procreative power.
The sitting position of the prolific father figure points to a loss
of the masculine life embodied in the upright stature of a man
on the move. Yet the loss is not without gains. Great power
comes from trials and penance endured in life, tribulations and

sacrifices ending in old-age infirmities. Joseph seen sitting in Genesis 50 (23ff.) is on the verge of dying, yet he is blessed with many great-grandchildren sitting on his lap. The implication is that masculine energy is rewarded if properly "spent." To be more precise, some of that strength must be relinquished or spent with ascetic intent if it is to bear fruit. This holds true especially in the reproductive domain. The preservation of male potency requires some loss of phallic power through abnegation. A procreator's strength cannot be preserved unless it is deserved. Masculine might is rewarded if obtained or enjoyed through some concession to the Lord – hence an oblation performed by means of ablation.

The oblatory gesture can take many anatomical forms. Men may gird up their loins or bend their knees before the Creator. Fasting has the disadvantage of reducing the strength of one's knees and the fat of one's flesh. All the same, only fasting can persuade the Lord to heal a lame body. Acts of devotion and self-denial will move God to set feeble limbs upright again and to make straight paths for the saintly followers who kneel before the Lord and bow down to his glory. All in all, the generative strength of a patriarch thrives on the oblation of an ablated joint – bending the knee, striking the hip, or tightening the belt around the loins. Life of the spirit coerces to obedience the lap, the loins, and the legs of a male creature otherwise hipped on the pleasures of procreation. Patriarchal figures must set part of their flesh aside with a view to making atonement for their sins.

Step 5: Stories of sacrifice bring us back to signs of great joy.

As already mentioned, limping Jacob became the eponymous founder of Israel immediately after the Lord had stricken his thigh and hip. The blow was struck in anticipation of the innumerable trials and sacrifices to be suffered by the chosen people issued from Jacob's loins. Life cannot be created or birth given without a woman's waters breaking. The same logic applies to the procreational process declined in the masculine gender. Man must gird up his loins and kneel before the Creator if he is to beget children and take healthy sons on his lap or knees. The womanly waters and the manly bones of human existence must be broken with sacrificial intent for the blessings of life to be secured. Life thrives on *la rupture des eaux/os*.

Foot and Shoe Fetishes: The Bright Side

Loins, knees, and heels endowed with manly strength stand for the good life. But if enjoyed without ever being renounced, the blessings they offer will lead to sin and great trials. The powerfully loined and the well heeled will be punished where they have sinned. They will be dispossessed of the manly powers vested in limbs and articulations of the lower body.

Our reading of Jacob's story and related imageries of the body points to the teachings of ascetic morality. Possessions will be lost by those who fail to relinquish them with sacrificial intent. But there are some intriguing anomalies and exceptions to this rule. Jacob, blessed with children sitting on his lap, enjoys the good life, yet he does not stand up with straight legs and feet of manly power. Rewards of the flesh do not extinguish signs of broken knees. A dispossession of manly strength seems positive after all, taking on meanings other than retribution or penance. The opposite qualification is equally valid: possessions of the body are not always pleasurable. Sartorial possessions of shoes and clothing can evoke the trials of life; men condemned to exile or war will thus cover their loins with girdles and feet with shoes. Why these surface inversions? Why should signs of loss accompany the blessings and sinful pleasures of life, and why should imageries of appropriation convey a life of troubles endured for reasons of punishment or sacrifice?

Answers can be found in the biblical lessons of foot and shoe imageries, starting with their joyful and sinful expressions. The sombre side (tribulational, sacrificial) of this symbolism will be explored in reticle 11.

THE JOYFUL

Our analysis begins with the least complex and most predictable scenario, which consists in the pleasurable aspects of shoes that are effectively worn and possessed. In the scriptures normal feet and shoes may be given positive connotations. One example of this lies in the firmness of leg and foot, which points to the wholesomeness and holiness of a healthy body. The opposite consists in lameness of the leg, an impairment that disqualifies men from the priestly office and is akin to a fool's unreliable proverb.[5] When endowed with strength, feet and footwear can also stand for the most precious possessions and blessings of life. They signify the kind of life enjoyed by those who have shoes made of iron and brass and whose feet are bathed in oil, butter, or blood of the enemy.[6] To the extent that they tread land and enter houses and habitat, feet, sandals, and shoes are signs of territorial occupation and property in general.[7] The larger the territory is, the more space and liberty there are for the feet; one's dominion comprises all things under one's feet and enemies or the earth reduced to a footstool.[8] Wealth and power can also take the shape of heavy-footed beasts mounted by those who tread, invade, or conquer land and its female occupant (e.g., the daughter of Israel). When associated with such animals, feet evoke the powerful claws, hooves, horns, and nails of a mighty, heavy-footed kingdom. They also evoke the horns of iron and hooves of bronze that rulers and armies can use to trample their enemies.[9] In keeping with this masculine imagery, the strength of a kingdom hinges on a kingly figure gifted with powerful loins and steady pins or legs. Similar tropes apply to the tent and solid walls of a kingdom, which is supported with stable pegs and words of wisdom driven deep.[10]

Men endowed with power and wealth are well heeled and well shod. The property they enjoy can be acquired by means of conquest. But it can also be obtained through proper negotiations and rituals of shoe transfer, hence through *foot of the fine* – formerly, in law, the last part of an acknowledgment of a title or transfer of land. We know from Ruth 4.7 that "in former times it was the custom in Israel, in matters of redemption or exchange, to confirm the transaction by one of the parties removing his sandal and giving it to the other" (see also Amos 2.6). Giving up one's shoe to someone purchasing property implied that the person to whom a piece of land was sold was the only one to have legal right to walk on the property.

Male property signified by the shoe includes the spaces men dwell in but also the women and wives they "possess." Since they penetrate sites of reproduction such as land and dwelling, feet have masculine attributions. The foot goes into the shoe like a man and his seed entering house, land, and womb. Given these connections, the shoe motif can be used to denote a transfer of wife, as in Ruth 4.5: "On the day you purchase the land from Naomi, you purchase Ruth the Moabitess also, the wife of the dead man, and so restore his name to his inheritance." The same imagery finds its way in Deuteronomy (25.5–12), which describes the levirate law and the fate of a man named House-of-the-Unshod. This is a man who never waited for dead men's shoes and refused to step into his brother's marital shoe and marry the wife of his childless brother. The man is guilty of shirking his brotherly duty. As a result, the woman to whom the man "owes levirate shall go up to him in the presence of the elders, take the sandal off his foot, spit in his face, and pronounce the following words, 'This is what we do to the man who does not restore his brother's house.'"

Feet serve to secure men's shoe-like possessions in the shape of land, houses, and women. Another motif that ties in directly with foot and shoe imagery is that of a door, gate, or entrance to a dwelling place, a temple, or a city. The door symbol signifies yet another enclosure entered by foot. Accordingly, both door and shoe motifs may appear when issues of property, marriage, and descent are settled, including Ruth's marriage.[11] Glad tidings can be brought to a nation or a man's house through someone stepping and knocking at a door and felicitous

Shoes for the Halitza ceremony, performed when
a man refuses to marry his brother's childless
widow. Leather, 15 x 28 cm. French, from Alsace,
18th C. Inv MAHJ 95.36.001. Photo: J.G. Berizzi.
Musée du Judaisme, Paris, France (Réunion
des Musées Nationaux / Art Resource, NY)

decisions made at the site of an entrance. We learn from the
Bible that a man stepping at a maiden's door and making a
friend of darkness outside her entrance can be the bridegroom's
friend. The messenger may be sent to ask the bride's hand, thus
heralding the feast of the model woman marrying the Son of
Man. When the door or gate is opened, manna can fall from
the sky and glory make its entrance into a holy site (Ps. 24.7,
78.23).

Property expressed through imageries of clothing, footwear,
and door entrance can take on positive connotations. Putting
on a shoe, putting the foot down, and "walking in" through a
door is to take possession, be it territorially or sexually. There
are circumstances, however, where a complete inversion of the
imagery changes nothing to its actual meaning. This brings us
to the riddle of lower-limb dispossessions to signify the blessings of life. The scene of a man uncovering his feet in bed is a
case in point. Here the shoe-removal metaphor is synonymous
with going to bed and having sex (Ruth 3.4); the act of unshoeing

a man implies barefoot nudity and foreplay. But the symbolism conveys something else as well. It suggests a symbolic substitution, replacing the shoe that denotes female property for a real woman taking the man in, letting him take possession of her and occupy her house and bosom. The latter substitution may also be obtained by a man washing his feet, accepting a woman's hospitality by uncovering his feet from traces of the land made of feminine clay (Song 5.3). The woman unshoes the man's feet cleansed of clay within her house. (Note that she can do this provided she is a woman fit for the man, a condition that becomes a problem when the man is a godly figure. After all, no one is fit to loosen Christ's shoes or to carry his sandals. Since divine shoes stand for God's immeasurable wealth, it follows that no woman can substitute her services and dwelling for the Lord's footwear. See Acts 13.25, Matt. 3.11.)

Scriptural imageries of dancing and lightness of foot point to a similar paradox. We have seen that powerful men are endowed with heavy feet, hooves, and legs made of weighty metal. By contrast, women's feet are light and "uplifted." Women can be so delicate that they never touch the soil or venture to set the soles of their feet on the ground (Deut. 28.56–66). But man can also derive joy from losing some of the might and weight attached to his shoes and feet. He too can seek pleasure by lifting his feet and experiencing the feminine lightness and tenderness of life. Given the right conditions, a text can thus make men's feet "like the hinds." It can give them the lightness of the roe roaming about the wild, the kind of agility that keeps the animal "from falling on the heights" (2 Sam. 22.34, Song 2.7–9). Feasts also give men the opportunity to lift their feet and leap with joy.

Many other blessings may follow from man's loss of heavy-footedness. Instead of trampling other lands and enemies underfoot, merry men may stamp their feet (Ezek. 25.6), dancing and putting their sandals and best foot forward. In lieu of dressing in the sackcloth of mourning or girding their loins for war, they can wrap themselves in gladness. Joyful men cover their feet with sandals (Song 7.1, Luke 15.22) and wrap their loins in linen and "girdles of merry dancing." No matter what David's prudish wife, Michal, may say, God himself may derive pleasure and praise from songs and Davidic dances performed in his presence. Likewise, acts of sincere repentance may be a prelude to festive events held in the name of the forgiving Father.[12]

But what does a man uncovering his feet in bed have in common with a man dancing and leaping with joy? The answer lies in the joyful surrender of gender attributes. It lies in man letting go his "manly" possessions so as to be at one with the other gender – the greatest possession. Belongings (shoes, heaviness of foot) that cover the body can be relinquished for joyful purposes provided they are metaphors for objects of desire (women), hence desires that possess the body. If objects in one's possession signify the likeness of one's desire, then it stands to reason that they should be relinquished when replaced by the "real thing" (removing the shoe when "entering the woman's door" and going into her bed). In the end, whether shod or unshod, light or heavy, feet can leave imprints of the good life on earth.

THE SINFUL

If lustful possessions and dispossessions of the body can procure pleasure, they can also lead to sin. Consider the evil consequences of objects of desire that are effectively owned – in other words, offences committed by those indulging in earthly possessions. Feet that stand for manly property are an invitation to sin. Feet made of iron can cause men to slip and slide into sinful behaviour, committing false steps that reflect their wickedness, to be scrutinized and measured by God himself.[13] Horsemen and fearsome armies "who march miles across country to seize the homes of others," plundering other lands and imposing their rule through sheer might, stand accused of following in the footsteps of the "sinful, he who makes his own strength his god" (Hab. 1.6–11). They use iron to conquer nations and to forge gods out of metal, fastening "the idol with nails to keep it steady" (Isa. 41.7, Jer. 10.4).

Nothing good can come of valour attributed to footmen walking tall on the path of war and striking the innocent while they are weak and staggering (Job 12.5). These men fall prey to temptations of sin and grow bold on the path of the flesh – the flesh with and over which they fight and make war. Their feet are impatient to shed blood as they conquer foreign countries, alien shoe-like property, and strange women too.[14] The wicked "are caught by the feet in the snare they set themselves" (Ps. 9.15; see also Prov. 29.6). Sin lies at their door (Gen. 4.3ff.).

But men can also behave sinfully by letting go of their shoe property, in exchange for pleasures of the flesh. This brings me to stories of men knocking at a woman's snare-like door, taking her out of her house or removing their shoes as they step into her house and bed. Take the story of an undefiled shepherdess whose heart was awakened to shameful desires by her beloved knocking at her door. The man had risen to his feet and walked to the maiden's house, as if to get a foot in her door. The Shulamite girl opened the door (of her house/heart/body) to her beloved and saw that he had turned his back and gone. She wandered out in the streets of the city in search of him. This caused her to lose her veil and all sense of shame. She behaved like a whore running *barefoot* into the streets after strange men and false gods (Jer. 2.25, 5.19).

The woman's dreams would have ended differently had she not answered the entreaties of her sweetheart. We know from the Song of Songs that a woman who behaves like a door runs the risk of being abandoned and put to shame by men guarding the ramparts of female virtue. A pure woman must abstain from leaving her home or removing her shoes when on the run. But she must also abstain from letting men enter and sully her dwelling. That is, she must resist the man's wish to take off his tunic, wash his bare feet, enter her body domestic, hence rush into the privacy of her unsullied home. She must not let her suitor thrust his hand through the hole in her door, causing "her bowels to move for him" and leaving her door and feet wide open to sexual attacks (Song 5.2ff., 8.9). In short, the lady should not behave like a portress inviting traitors to her house.

She should not follow the example of the "damsel" who made the Lord indignant by admitting Peter to a palace of false priests, a place of whorish infidelity (John 18.16). Nor should she lie down with wicked men assembled at the door of the tabernacle of the congregation (1 Sam. 2.22).

A man of virtue should not let himself be attracted by a woman of the kind that "opens her door" to neighbours passing by her house. The woman behaves like a harlot who refuses to live in seclusion. Suitors knocking at her door get off on the wrong foot/path. Neighbours who flatter her vanity are setting themselves a trap, especially if the lady answers their call. Because misguided, their foolish feet will get caught in the net of sin and burn in hell. They should withdraw their feet from this woman's house and ensure that they are not found *unshod in a house of sin*. Feet are inclined to behave foolishly and shamefully and should be covered accordingly. As with other less honourable parts of the body, they have to be clothed with care. "So our improper parts get decorated in a way that our proper parts do not need. God has arranged the body so that more dignity is given to the parts which are without it" (1 Cor. 12.22–24). A man uncovering himself as he steps into a woman's interior loses all sense of shame.

Little imagination is needed to put this foot and door imagery on a firm sexual footing. A man walking and knocking at a woman's door may end up knocking her up and making her pregnant. He can be suspected of befriending a woman's entrance, dark works of iron (1 Kings 6.7), and the cult of false fathers and related acts of lewdness performed behind closed doors and in darkness: "Behind door and doorpost you have set up your sign. Yes, far removed from me, you unroll your bedding, climb into it and spread it wide. You have struck a pact with those whose bed you love, whoring with them often with your eyes on the sacred symbol" (Isa. 57.8). A man's loss of heart and soul lurks at a woman's door turned into a harlot's snare (Job 31.9, Prov. 7.23). Days of troubled darkness await suitors appearing at a door giving entrance to the house/body/womb of a maiden playing (footsie) with desires of the flesh.

Men removing their shoes in a house of sin make light of their moral duties. The sinful lighten themselves of their moral obligations in yet another way: by engaging in acts of festive dancing performed through lightness of foot. While expressing joy, dancing may be symptomatic of frivolous behaviour on the part of hoofers who pay no heed to warnings of danger. The men in question indulge in jumping and feasting as opposed to kneeling, praying, and fasting. Great danger lies in feasting without fasting. Sin comes from feet uplifted above the ground, or the foot of pride evoking not only men of power and arrogance but also their delicate wives whose soles never venture to touch the ground (Deut. 28.56–66, Ps. 36.11). Men should stay away from women caught walking "with their heads held high ... tinkling the bangles on their feet" (Isa. 3.16). The warning evokes the sinful act of Herod who murdered John the Baptist in exchange for a dance performed by the impious daughter of Herodias (Matt. 14.6, Mark 6.21f.). It also brings back the sad memory of Jephthah's daughter, a virgin child who came out of her father's house to welcome the return of her father with dance and music. The father was a man of war who had taken revenge on his enemies, trampling them underfoot and humbling them before the Israelites. The feet of daughter and father conjoin themes of female dancing with imageries of manly power. The scene offers joys that are so worldly and cumulative as to end in tragedy. By leaving the house, the girl forced her father to offer his only daughter's life and virginity in sacrifice according to his vow to Yahweh (offering the first person coming out his house in exchange for his victory over the Ammonites) (Judg. 11.31).

While they can mark joyful moments, signs of feet, shoes, and doors can be perverted and lapse into crime and evil. Be they unshod in the house of a barefoot woman or shod and marching on the path of war, feet can leave imprints of sinful behaviour. Traces of immorality can also be found in scenes of either heavy-footed men at war, or their light-footed counterparts leaping in the company of delicate women dancing with sandals and foot ornaments.

Signs of reproachable desire can be conveyed through enviable possessions and endowments of the foot. But these possessions can also be removed to make room for delicate feet and pleasures of the womanly flesh. The end result is the same; sin triumphs in both the cases.

Foot and Shoe Festishes: The Dark Side

THE TRIBULATIONAL

What happens to women who open the door to male feet slipping into sin? What is to become of men who show the foot of pride? Will feet that wander from the path of God and run to mischief go unpunished? Answers to these questions can be found in tribulational usages of foot, shoe, and door motifs. Lower body and door entrance images evoking scenes of sinful pleasures can be adjusted to illustrate the fearsome implications of Eros let loose. That is, they can be turned against men and women abusing the powers over life (hoarding, begetting) and death (conquering, killing, punishing).

Retribution can nonetheless take one of two opposite forms. The most predictable form consists in objects of pleasure being taken away (scenario 1). Feet, legs, shoes, and doors can be adapted to stories of divine wrath expressed through stories of dispossession and dismemberment. But judgment is not always carried out through signs of loss and deprivation. Retribution can also be imposed through stories of harmful possessions (scenario 2). Belongings that constitute or cover the body can be used to emphasize the self-destructive powers of worldly endowments, assets that will neither simply come together (through metonymy) nor merely "stand for" one another (through metaphor).

SCENARIO I: PUNISHMENT
THROUGH DISPOSSESSION

The first scenario consists in God punishing the wicked by depriving them of what they desire the most, or metaphorical expressions thereof. Scriptural scenes of retributive dispossession applied to foot, shoe, and door imageries are many. The well-heeled and the well-shod incurring the wrath of God are typically stripped of their shoes worn out in exile (Josh. 9.13). The Almighty divests them of the wealth covering their bodies and protecting them from poverty and shame. Recall that members of a well-heeled nation are prone to behave like whores running barefoot after strange gods; accordingly, they deserve to lose their loved ones, and hence remove their shoes while in mourning. They can expect to be stripped of all their precious possessions and join the footworn and barefooted have-nots of a fallen humanity (Isa. 20.1–4, Ezek. 24.16ff.). Likewise, God may sentence a woman arrayed in wealth to be divested of all the ornaments (chains, belts) worn around her head, neck, waist, and feet. She can expect to be unshod, her heels to be made bare (Jer. 13.22), and her shoes to wear out in exile. Yet were they to relinquish all forms of wealth, the lady and her people might not lose their precious shoe-property as they leave and mourn a place fit for princes.

SCENARIO 2: HARMFUL POSSESSIONS

Dispossession can spell ruin. The alternative scenario, however, can be equally devastating. God may chastise the wicked by letting them keep endowments that ultimately bring harm even to those who possess them. Take, for instance, the self-inflicted sufferings imposed upon legs and shoes of iron. As already mentioned, redemption is never granted to men who take pride in their natural strength and instruments of might and self-idolatry (e.g., idols, chariots, horns, and hooves of iron) – on the contrary. A kingpin who lets his iron constitution go to his head can fall to his death with a nail-peg from his enemy's tent driven into his head. Images and works of iron turn against those who use them inappropriately. Forces of violence rebound on men of war.

Instead of serving as protection, the firmness and strength of a man's legs and feet may be destructive of his own life. This is in keeping with the fact that men of power are prone to waste the soft flesh they seek to possess and reproduce. The imageries of legs and feet of iron indulging in carnal pleasures and weaknesses of the flesh are so full of contradictions that they offer constant invitation to disaster. Biblical symbolism applied to the anatomy of power can serve to reinforce this point. According to Daniel, the legs of a powerful man and his kingdom are firm and solid, as if made of iron. Those legs, however, are also made of flesh, or the chthonian analogue thereof, the fragile clay used in ancient times for sealing doors. The good life thus unites the male and the female. Accordingly, works of metal go hand in hand with the multiplication of women and the cult of beautiful goddesses; iron keeps company with the lustful life that makes idolatrous men soft. Strong-limbed men may wish to secure pleasures of the clayish flesh, yet the synthesis of iron and clay is full of problems. Because of the Fall, the two ingredients of life on earth do not mix well. Though the two substances may be combined in the human body, they cannot hold together any more than iron will blend with earthenware. Iron breaks the human vessel.[15] Paradoxically, men's iron constitution leads to consumption and destruction of the flesh. An excess of virility converges on a wastage of life, the desolation of land and sky filled with iron and bronze (Lev. 26.19, Deut. 28.23).

Excessive strength attached to lower limbs spells ruin for the mighty. We just saw that might that does not last can take the form of iron legs. But it can also take the shape of uncloven feet. Beasts that exalt their horns of power above the reign of the Lord and are never so weak as to "part the hoof" are unclean creatures. Biblically speaking, quadrupeds that ruminate and have cloven feet are clean and edible. Horses are unclean, for they do not have cloven feet. To the horses' impure

status can be added their close association with the tools of
war and the solid feet and weapons of the flint-hearted men
who mount the beasts. The animals also evoke men's tendency
to nail their self-idolatrous colours to the wall or the mast of
their own "undivided" house. The animals' uncloven feet also
evoke men's longing for houses that are powerful and therefore
never "divided." But while they may be uncloven and are associated with the works of solid iron, creatures of wrathful disposition nonetheless bring in their wake evocations of a nation's
house divided, scattered and turned into a military forge. They
conjure up visions of inexorable trampling and wheels of fire
whirling over the land.

The iron motif is closely tied to stories of destruction, dispossession, tyranny, and war. Men whose feet are quick to shed
the blood of wrath like horses running to war are heading for
death. They shall be trampled by horsemen and heavy-footed
beasts hailing from abroad. The same fate awaits iron-souled
men who succumb to religious whoredom and carve or cast
works of heathenry with their own hands. They too shall join
their victims and be trampled underfoot. Their arrogant feet
bent on crushing other nations shall be put in irons. They shall
lick the dust from the feet of their conquerors. They shall yield
to the rule of iron-fisted masters, cruel men of their own kind
yet more powerful than they, who torture the feet of their
victims with fetters and put their necks in irons.[16]

Men who possess (horses wearing) iron will be trapped into
iron. Likewise, soldiers wearing girdles and shoes may turn
from warriors conquering foreign lands into expatriates fleeing
in exile, never to own land of their own. This means that the
girdles and shoes men wear and the saddles they mount may
be indices of a self-inflicted exodus, not articles of the soldierly
apparel; the same imagery turns men of war into the landless
and the destitute. Not even the horses they mount can come to
their rescue. Wicked men are bound to suffer the consequences
of their own horses wearing iron shoes, an imagery that suggests one of two things. The animal may be on the verge of
trampling a sacred space, and thus punishment will ensue.[17]

Uncouth outsiders keeping and tightening their girdles about their loins and shoes on their (horses') feet are fated to be conquered by forces more powerful than they, enemies entering and spoiling their land, houses, and holy sites (Isa. 5.27). Alternatively, the ground on which the beast stands is not exactly holy, an imagery that points to the story of a people deported and dying on their feet while wandering in foreign lands. Such is the fate of an attractive daughter/nation being spoiled, raped, and booted out, expelled from the doors and gates of her own house and homeland. Victims will dress accordingly, for a journey in exile, away from the land of Canaan, in the direction of bondage. They will wear sandals and tighten their belts or girdles, never again to hoof their own land. God will force men to wear their attire as long as they remain in exile, never to own the places they tread nor to find rest for the soles of their swollen feet. Exodus conveyed through the wearing of the shoe is where the shoe pinches the most.

In summary, while sinners are fated to be stripped of their precious possessions, retribution can also take the form of turning precious possessions into objects of self-inflicted harm. When this scenario prevails, assets are no longer relinquished for what they stand for, as in scenes of sinful pleasures (shoes removed in a harlot's shoe-like house). Rather, the object possessed becomes a condemnation in disguise, forcing the possessor to forego all the things the object normally stands for. Trials and sufferings inflicted on the sinful thus involve people wearing shoes because they have been exiled and dispossessed of their homes and lands, thrown out of their gateways and doors (Josh. 9.5).

As we are about to see, the same paradox applies to tribulations suffered in the domain of sex and reproduction. On the one hand, bodily afflictions expressed through foot, shoe, and door symbolism can take the shape of anatomical losses, as in imageries of dismemberment and disembowelment (scenario 3). On the other hand, chastisement can also come in the form of signs of morbid occupation and self-defilement; the land possessed becomes a burial ground, and the power covering the body turns into sheer filth (scenario 4). We turn to these opposite expressions of tribulation, starting with losses suffered through dismemberment and disembowelment.

SCENARIO 3: BODILY LOSSES

Evil men can be penalized through anatomical losses conveyed through evacuation, expulsion, decomposition, or dismemberment. Judgment can cause the body to suffer great losses through evacuation of the bowels and related signs of all hell let loose – people expelled from their land, a seastorm bursting from the jaws of hell, and blood pouring from the human body. Men and nations indulging in sin are bound to be expelled from their land. As with the Benjamites, men courting harlots and engaging in acts of infidelity expose themselves to God's wrath. To use the language of reproduction, the womb that the unfaithful seek to defile may swell up with signs of death and turn against the unfaithful. The womb that gives life thus turns into the womb of a whorish sea no longer shut up behind doors.[18] Wicked men lay themselves open to either a flood of tribulations bursting from the jaws of hell or the Almighty coming out of his dwelling to punish the sinful for their crimes (while protecting his followers hiding behind closed doors, preferably in a sanctuary) (Neh. 6.10, Isa. 26.20–21). Signs of lethal bursting and evacuation include the swelling of feet and the pouring of blood. As in the story of a harlot named Rahab, women going to the streets are responsible for the shedding of family blood (Josh. 2.19).

The lap motif can be used to express similar trials attained through *expulsion*. We know from Nehemiah that those living in the lap of luxury are condemned to shake in their shoes for fear of losing their comfortable home – of being shaken out of God's lap: "Then I shook out the lap of my gown with the words, 'May God do this, and shake out of his house and property any man who does not keep this promise'" (Neh. 5.13). By implication, exile awaits a man who takes off his tunic in a house of ill repute. The man succumbs to a woman of nocturnal pleasures. He falls prey to a woman no longer covered with the cloak of purity and confining herself to her house, shutting the door to outsiders.

Expulsion is also part of the nightmare of the harlot who has lost her hymenean veil. Men stepping and knocking at this woman's door can be suspected of having only one thing in mind, to abuse the "strange woman" dwelling inside, leaving her dead at the door, like the harlot who accompanied the Levite sojourning in the land of Gibeah (Judg. 19). This imagery points to the vulnerability of the female cavity of life and the sombre fate of men caught spoiling it. Traces of manly penetration jeopardize both the purity and the life of a woman's heart. Men are guilty of violating the maiden's well-protected "enclosure" (*enceinte* in French, also for "pregnant"). They can expect to suffer a fate similar to that of the damsel whose "tokens of virginity have not been found."[19] Death and the Lord will break through the walls of their own house like thieves. Breaches of faith result in death entering the walls of a nation lusting after the good life on earth.[20]

Men waiting at a harlot's door are bound to spoil the woman's life as well as their own. Signs of mortal expulsion can generate stories of expatriation, people expelled from their land like blood gushing out of their body. Death by "physical evacuation" may also lead to the decomposition of limbs and members of the body, hence the sufferings of emasculation or dismemberment. But how can whoredom be punished through emasculative usages of the foot and the shoe imagery? The answer lies again in the afflictions of a man who yields to temptation and sets foot in the house of a harlot. Since his lower limbs are slipping into sin, they deserve to be severed from the body. A man who betrays his master should have his feet cut off. He should be deprived of the ability to move and he should go to his grave with his feet bound up.[21] Likewise, someone caught "playing footsie" with idols that cannot walk is guilty of letting his feet slide into sin. Logically, the man stands to lose his footing; eye for eye, tooth for tooth, hand for hand, foot for foot

(Ps. 115.7, Exod. 21.25). "*Qui prend son pied risque de le perdre…*" This is to say that "if your hand or your foot should cause you to sin, cut it off and throw it away; it is better for you to enter into life crippled or lame than to have two hands or two feet and be thrown into eternal fire" (Matt. 18.8).

SCENARIO 4: MORBID OCCUPATION
AND SELF-DEFILEMENT

Judgment can be expressed through uterine expulsion and anatomical dismemberment. Losses are suffered by those who fail to heed teachings of the Lord. But judgment does not always entail signs of dispossession. Tribulation can also take the shape of signs of morbid occupation and noxious possession. This brings us to the pains of self-abuse or self-abasement resulting from acts of penetration. To begin with, entering a house of sin may result in death by interment, a far cry from acts of domestic reproduction and land occupation. Men's inclination to lean over female flesh, putting their feet up and lying down with women in bed (Gen. 49.33), amounts to a death wish, the wish to have at least one foot in the grave and fall into a whore's bottomless pit. Like the image of jealousy sighted at the entrance of Jerusalem, a whore sitting at the door of her house and attracting passersby is bound to lead men "in heat" to burning in Sheol. Depravation on earth engenders deprivation in the womb of hell.[22]

We have seen that a man caught unshoeing his feet and stepping into a harlot's house deserves to be unshod and rejected. God will see to it that he is purged or expelled from the shoe and the woman or land it stands for. These trials are reminiscent of the misfortunes of unshod David fleeing from the son who stole both his wealth and his wives. Readers should also recall the barefooted man surnamed House-of-the-Unshod (Deut. 25.9f.). This is the man who wasted his brother's name and property by refusing to fill his marital shoe. He declined to offer his procreative seed for the preservation of his deceased brother's name. The offence cost him his concubines, his house, and his shoes. The same would have happened to Boaz had he not recognized Ruth's claim to redemption by her kinfolk.

But instead of being unshod, feet caught in the act of entering shoe-property, women, and strange lands can be condemned to contain or wrap themselves even more. In lieu of securing reproductive gains, feet caught entering a harlot's house may end up covering themselves with filth – female earth, clay and flesh gone sterile and offensive. Since a sinner is prone to cover himself with filth, his feet might well be literally defiled, sullied by his own waste matter (Judg. 3.24). This excremental language harks back to Saul losing his cloak along with the dignity of the royal cloak and mace placed between his legs as a sign of territorial authority (Gen. 49.10, 1 Sam. 24.4). The king soiled both his feet and his reputation as he fell at the hand of a contender to his throne. The kingdom and sceptre or mace held between Saul's feet were lost to young David whom God elected to step into the king's shoes. Given improper behaviour, the reproductive powers of female soil and phallic leg and mace will quickly perish and spoil through signs of decay, excretion, and waste.

Even the food of life can be corrupted when undeserved. Expressions of defilement applied to feet, food, and reproduction point to the daughter of Israel who ate her husband's flesh along with the arrogant filth of her children emerging from between her feet. The lady became so vile in the eyes of God that he plucked her out from the land of Canaan. Her troubles in exile were such that her "delicate" heart and feet never found rest. The house, the woman, and her children perished and went to waste. An impure womb brings nothing but sorrow, misery, and filth (Deut. 49–68, Job 3.10).

A few words should be said about the *tribulational aspects of dancing*. Can festive movements of the body signify trials and sufferings inflicted upon the sinful? Again the answer to this question goes in one of two directions. The first route involves anatomical losses. Feet and heels uplifted with joy may be suspected of rising against the Lord. Accordingly, they belong to haughty spirits destined to be lowered: "Pride goes before destruction, a haughty spirit before a fall" (Prov. 16.18). Unless tempered by remorse, men's leaping with joy and pride exposes their knees and feet to the wrath of the Lord. God smites the joints and limbs of those caught dancing and stumbling in religious darkness (Jer. 13.16).[23] The Father can summon them to fall on their hands and knees and to wait on him hand and foot. His anger is such that he may place his feet on the necks of the haughty and break them (Josh. 10.24). If dancing is performed in honour of the calf of biblical times (Exod. 32.19), the trials of judgment are all the more inevitable. A nation in raptures over an idol made of flesh is as vulnerable as the calf itself. Like the animal, unfaithful daughters shall suffer the hardships of abduction, a destiny that harks back to the violent theft of the dancing virgins of the ill-fated people of Ja'besh-Gil'ead and the story of a people's inheritance lost. Women dancing or walking with their heads held high shall go naked and be stripped of the bangles adorning their feet (Isa. 3.16ff.).

But scenes of dancing and suffering can take another direction. The dance may continue, but the narrative can transform the festive moment into a sombre reflection of what it stands for. God can thus lead merry men leaping with joy into the Dance of Death (Isa. 22.2, 12f.). He can let them dance like the demonic satyr or the worthless goat, with the implication that they will march into a land of desolation.[24] Idol worshippers making merry will thus be cast into the realm of the dead, a wasteland ruled by gods who have limbs of gold and silver but can neither touch nor walk, let alone dance (Ps. 115.7).

To sum up, men whose feet are attracted by idolatry, whoredom, and the tools and spoils of war can be punished by being dispossessed of what they value the most. They may be stripped of their shoe-like property and lower limbs. Feet caught sliding into sin may be unshod, thrown out from home, the motherland – hence the female shoe-property they so carelessly possessed. Death bursts out from hell and breaks into evil houses, dismembering the wicked nation. Sufferings sent from the womb of Sheol expel the sinful from their land and home, putting an end to all the blessings that lie in the lap of luxury, life wrapped in girdles of merry dancing.

But punishment can be dealt through other means. Instead of dispossessing men of what they treasure the most, God may see to it that men's possessions preclude everything they normally stand for. Every object of desire is then turned into a condemnation in disguise, an exercise in self-abasement and self-destruction. In the end powers acquired and sufferings imposed upon others turn against the wicked. Punishment comes from the much-vaunted possessions of iron, metal works that break up and destroy the clay-flesh, the houses, and the empires that men of iron strive to possess. Men of iron are crushed and trampled by uncloven feet and weapons mightier than their own. Warriors are fated to keep their shoes and girdles on, wearing the same signs as the people they conquer and force into exile. Similarly, feet may be allowed to step into a harlot's house, but they do so at a great cost to men's reproductive powers. The house and womb they penetrate and possess turn into signs of great morbidity – signs of the grave and the bottomless pit. Feet and bodies well-heeled and well-shod are fated to end up covering themselves with waste and filth. And while wicked feet may continue to dance, they are doomed to do so in a land of desolation and immobility, a wasteland spoiled by the Dance of Death.

THE SACRIFICIAL

Feet, shoes, and doors can be endowed with blessings and powers (over life and death) that can turn abusive and sinful and lead to tribulational usages tied to visions of judgment and wrath. But tribulations can also be transformed into signs of sacrifice. If fully assumed and offered to God, ordeals can translate into self-sacrificial offerings that will secure God's mercy and providence. Once again acts of propitiation and redemption achieved through ascetic behaviour can be expressed through foot-shoe-door symbolism, signs that lead us back to imageries of joy and prosperity that are truly deserved.

I should reiterate here that signs of the ascetic life do not involve a simple choice between the possession or dispossession of objects of desire wrought in the shape of a shoe, a foot, or a door. Rather, the issue is whether or not the body language and related paraphernalia are informed by the sacrificial disposition. For instance, shoes can be worn by the wealthy treading their own land. But they can also be worn by those who take the opposite direction, away from home. The latter option corresponds to what homeless messengers do as part of the sacrifices they must endure abroad while spreading the word of the Lord. Footwear can belong to men of God who relinquish all the good things that shoes otherwise stand for.

Similarly, shoes that are removed can mean different things. A bare-footed figure is as ambiguous as a bare-headed person. The implied nudity may result from the enjoyment of comfort at home, where shoes and hats are conventionally removed. But nakedness can also attest to a life of poverty and renunciation. The same paradox applies to the girdle motif. Biblically speaking, a man wearing a girdle about his loins can be carting around wealth in his reproductive lap, the wealth that girdles of merry dancing stand for and over which men normally rejoice. But the same man can also be preparing himself for hard times to come. He may be girdled without being vested with the blessings normally wrapped in it. In short, the real question lies in whether or not the text is wearing or removing the shoe or the girdle as an expression of the pleasure principle or an effect of the ascetic ministry of the Lord.

Lessons of sacrificial behaviour thus lend themselves to bimodal expressions. They can be communicated through either imageries of possession or of dispossession. Stories of dispossession are used more frequently. Biblical references to maiming are instructive in this regard. In the scriptures, bodily infirmity is reproachable in the sense of disqualifying a man for the priestly office; an offering pleases the Lord only if it is without blemish, a worthy gift to the Highest (Lev. 21.18, Deut. 15.21, Mal. 1.9, 13). People disabled with sacrificial intent, however, differ from those lamed by misfortune or punishment in that they turn infirmity to advantage. To quote the Letter to the

Hebrews (12.11–13), "Any punishment is most painful at the time, and far from pleasant; but later, in those on whom it has been used, it bears fruit in peace and goodness. So hold up your limp arms and steady your trembling knees and smooth out the path your tread; then the injured limb will not be wrenched, it will grow strong again."

Maiming can bring great rewards. Come the Millennium and the reign of the Lamb slain, blessings shall be bestowed upon the humble falling on their knees before the Lord. The Lord shall call them to his love feast and make them rise to heaven (Luke 14.13). The maimed shall be made whole and the lame shall be healed; they shall stand up, walk straight, and leap like the hart.[25] Saints and martyrs shall be resurrected, just like their Messiah portrayed with hands and feet pierced (Ps. 22.16, Luke 24.39f.). In the end, it is God the Healer who cures diseases and infirmities of the foot, not the physician (2 Chron. 16.12).

Men must submit to signs of lameness or infirmity if they wish to follow in the steps of the anointed one and attain wholesome unity in Christ's body. An oblation performed through an ablative operation and the surrendering of life brings everlasting rewards. By "putting off some flesh," men can covenant with God for redemption. Illustrations of this principle can be found in the foreskin cast at the feet of Moses, or Christ's pierced hands and feet substituted for the entire body of his flesh and bones. Alternatively, followers of God may choose to cut off the offensive hand or foot in order to secure the wholesomeness of body. This imagery of limbs torn from other limbs can be used to rearticulate the moral of self-denial. This is to say that signs of emasculation and dismemberment may inspire feelings other than the fear of death. Life aborted and unmanned is not always gruesome. The maiming imagery can serve to counter the forces that turn men away from the life of the spirit and rewards of the hereafter. Maiming can lead to a wholesome afterlife on the condition that the body is impaired with sacrificial intent.

Redemption requires cutting operations (severance, circumcision) modelled on the rule of *self-denial by ablation*. This entails the kind of sacrifice performed by a messiah whose role it is to

set the example for his own followers. Salvation thus lies in the
wastage of a Saviour resembling Christ the Rock. Consider the
refuge of a rock. While the king of Babylon was gazing at the
statue of the human empire through time, a stone broke away,
though untouched by any hand. The piece of rock struck the
statue's feet, made partly of iron and partly of clay, shattering
the entire monument and scattering the pieces like chaff to the
four winds. The stone, however, grew into a mountain filling
the whole earth, forming the indestructible kingdom of God
(Dan. 2.33–45). The point to be made here concerns the recov-
ery of a stone cast away through ablation. Given its sacrificial
inspiration, this form of decomposition can feed into the life of
the spirit, using the lowly cast-off condition as a stepping stone
to life in the hereafter. The rock motif can become wholesome
and all powerful by virtue of some Great Divide, an original
fault or rift inscribed in human nature and reparable through
the intervention of a Redeemer capable of mending the breach.
Again, the strength of the imagery of rock (and iron feet) lies
in its sacrificial usage.

A script expressing the teachings of sacrifice could also invite
followers of God to renounce all works of metal, which is
another way to evoke the shattering of iron. This would make
sense given that the works and the house of the true Lord are
built without idols of loud metal. God's holy ministry is incompatible with the use of tools associated with the iron works of
war. Unlike the harlot, the lady of wisdom puts no value in
metals of the underworld (Job 28.2, 12). Those who fail to
follow her example shall be excluded from the city of the heavenly king, a site more awesome than all the metal goods and
vain constructions of the human hand.

Alternatively, signs of an ascetic spirit can be conveyed through
a divestment of footwear. Sacrifice can be obtained by removing
shoes and clay from one's feet. Ablutionary precautions can
also take the form of consecrated ram blood placed on the right
foot, a priestly ritual where redemption is attained by pouring
the blood of a slain lamb. The notion that the humble should
uncover or wash their feet when mourning and praying to the

Lord in a holy place is a reminder that comforts of the body must be given up if greater joys are to be deserved.[26] While stripped of their footwear, the humble should stress how unfit they are to wear God's shoes, let alone set foot in his garden. They should show reverence for their master by kissing, falling at, or sitting at his feet.[27] Alternatively, the faithful can undertake a barefooted journey into exile, towards acts of redemption achieved through missionary zeal – the opposite of well-heeled men who cling to their shoe-like motherland and fail to renounce worldly property (1 Kings 2.5, Matt. 10.9f.).

Sacrifice can be performed through acts of self-deprivation. A demonstration of ascetic virtue, however, can be achieved by holding on to certain possessions. Keeping the shoes on is an acceptable option. But there is a condition to be met. Feet can step inside the shoes as long as they keep outside of and away from the pleasures they stand for (home, land, women, etc.). Thus piety and humility can be obtained by wearing one's shoes while suffering in exile. The scene is ritualized on the occasion of Passover celebrations (Exod. 12.11), denoting either vigilance and watchfulness at home, or "eagerness to spread the gospel of peace" in foreign lands.[28] Given the ordeals they suffer throughout their journey to the Promised Land, the faithful can expect God to prevent their feet from swelling and their shoes from wearing out.[29] Men will be rewarded provided their feet never falter or deviate from the path of God who enlightens and safeguards their steps. They will be blessed on condition that they walk firmly in the Lord's footsteps, away from all imprints of man and woman falling into sin.[30]

In the same vein, the man who steps into his brother's marital shoes is assuming a burden. The man seems not to be relinquishing anything: on the contrary, he is inheriting his brother's possessions. But readers of Exodus and Deuteronomy are reminded that the levirate marriage is defined mostly as a fraternal duty, not an inheritance right. The Mosaic rule meant that when an Israelite died without having had sons, his brother (on the father's side) was under the obligation of marrying the widow, raising the children issued from such marriage, and then transmitting

the deceased man's property and name to the first son born from
the levirate union. The man who married the widow did not
inherit his brother's wealth. Rather the man who stepped into
his brother's shoes protected the deceased man's possessions and
assumed the burden of someone else's reproductive responsibil-
ities.[31] If he was unwilling to assume this responsibility, the
widow was entitled to humiliate him in public by plucking off
his shoe and spitting in his face.

THE METAPHORICAL: TAKING A STEP BACK

When enduring the trials of judgment or sacrifice, men and
women are predictably deprived of what they value the most.
The condition applies to the men of God in missionary exile,
men who must relinquish possessions of this world if they are
to deserve them. Abnegation of this kind is not a negation of
all human desires. Rather, it is a statement to the effect that
possessions cannot be obtained without some cost or payment.
But why should the man foregoing the blessings of life on earth
be allowed in some cases to keep some of his possessions – his
footwear, for instance? Should he not be divested of all belongings, signs of shoe property included? Why doesn't the Bible
systematically uncover the feet and require the man to suffer
total deprivation and roam barefoot in exile?

These questions can be answered through a better understanding of the language of metaphor. If in exile, the man's
shoes resemble the feminine wife-house-land possessions he has
lost. The footwear, however, pales into insignificance beside his
former belongings. Shoes worn in foreign lands can thus fulfil
an ascetic function by virtue of what they preclude, namely, an
effective possession of what the shoe stands for: property, land,
and women. When worn with ascetic intent, the shoe metaphor
acts as an "unpleasurable possession." It serves as a reminder
of what the man is lacking and yearns to recover through
exemplary behaviour. While a meagre possession, the bare min-
imum of a person living on a shoestring, the man's shoes simulate what he longs and dreams for: re-entering the mother

church, taking possession of the Promised Land, and securing the powers to reproduce without further trials. Shoes worn in exile are a man's most precious be-longings.

To possess is to enjoy a gain. To dispossess (a RH withdrawal action?) is to suffer a loss. One principle seems to denote limbic pleasure. The other looks like an expression of prefrontal judgment, a concession to signs of law and morality. So we commonly think. The previous analyses have shown things to be slightly more complex. Semiotic connections assigned to recurring imageries can be steered in different directions, depending on the overall configuration of sign attentions and the dominant effects being pursued.

Our sign-malleability thesis has been illustrated at length, using biblical imageries of shoes, heels, feet, knees, hips and loins. The argument has been extended to stories of possession and dispossession applied to anatomical and sartorial motifs. Marks of appropriation and removal applied to limbs and clothing were shown to serve variable interests, be they joyful, sinful, tribulational, or sacrificial. This phenomenon of the malleability of signs, however, has little to do with the rule of freeplay in semiosis. In reality, freeplay is merely the end product of the two-sided role of metaphor. Signs acting as metaphor (e.g., shoes) link up with the signs they metaphorize (e.g., homeland property) through one of the two strategies. The first strategy is of a syncretic nature. It requires that the signifier simply accompany the sign that is signified, both being possessed simultaneously. Shoes can thus be worn by feet occupying the land they own, be it deservedly (the joyful scenario) or not (the sinful scenario). Alternatively, both signs may be denied; shoes will be removed when feet are on the move and in exile for reasons of either punishment (the tribulational scenario) or missionary activity (the sacrificial scenario). Signs linked in this fashion are like birds of a feather that flock together, signs flying in or out of our sight at the same time.

The second strategy is radically different. It consists in metaphor used diacritically, as an act of simulation and yearning – a pale reflection of what it resembles but does not immediately deliver. Shoes can thus be worn to signify a journey to the Promised Land or exile from it, which amounts to the same thing: a longing for what the shoe stands for but does not surrender. Well-shod travellers are blessed or punished with yearnings for the land. Logically, this shoe-possession (used diacritically) may be "removed," "withdrawn," or "lost" when the blessing is granted. Signs linked in this fashion are like birds of good or ill omen and the meaningful desires or fears they stand for: they are kindred spirits not obliged to meet. The birds may leave the scene as soon as what they stand for is put on stage.

Spikes and the Motions of Desire

An intriguing paradox emerges from our interpretive findings concerning biblical imageries of feet, shoes, and doors. Asceticism involves signs of deprivation, yet possessions covering the body may be needed to signify what is longed for (higher rewards), typically by way of metaphor. When transposed to the sexual domain, the implication is that a male figure behaving virtuously can retain some feminine possessions or tokens thereof, using them as indications of what is longed for and worthy of a sacrificial offering. Paradoxically, these "ascetic possessions" are not without erotic value.

This brings us back to a critique of the conventional wisdom that pits ethics of self-discipline against Eros, a dualism that oversimplifies the workings of prefrontal judgment and limbic sentiment. Codes of self-denial are by no means purely negative or proscriptive. After all, rewards must be secured for the sufferings endured by models of virtue, be they virgins, widows, the faithful, saints, or martyrs. Promises made to the virtuous must be as alluring as the pleasures renounced, if not more enticing than the joys of sin. To use the Bible's gendered imagery, the model woman is entitled to live in the hope of receiving all the blessings that the harlot fails to preserve because she never deserved. In the end concessions to the flesh will have to be made, if only by way of sublimation. Losses suffered through sacrifice represent only one side of the ascetic equation. The other side covers the virtuous body with signs of *erotic belongings*, traces of the model woman's desire to be possessed

by the man of God. The model woman is both entitled and
expected to express her hopes of being united with her Shepherd
Lover. She can do so through a simulation of the union itself.

Descriptions of the womanly ramparts of virtue can serve to
drive this argument home. In the scriptures, protections for a
woman's chaste body are typically constructed in the masculine
gender, in the likeness of soldiers guarding a woman's purity.
When virtue prevails, masculine obstructions are part and
parcel of the womanly anatomy and hers alone to possess.
Recall that a woman eager to show her virtue shuts herself
behind closed doors. Her gesture is condoned provided her
intent is to please her Lord, as opposed to taking advantage of
pleasures enjoyed in the privacy of her dwelling. A woman pre-
serves her chastity by keeping her door shut and dwelling in a
house made barren through sacrifice. Yet for all that she is not
fully isolated from signs of masculinity. She is still expected to
ask God to guard the door of her lips (Ps. 141.3) and men to
guard the ramparts of her virtue. Like the Shulamite girl, a
model woman can close her womb for reasons of spiritual
health on condition that she equips herself with the manly
strength of an impenetrable wall, an image of virtue that con-
notes the opposite of a knocking-shop's movable door. To quote
the Song of Solomon (8.9), "If [our sister] is a rampart, on the
crest we will build a battlement of silver; if she is a door, we
will board her up with planks of cedar. I am a wall, and my
breasts represent its towers."

Masculine ramparts play a key role in preserving a woman's
virtue. But the same results can be achieved through opposite
means. A woman can hold on to her virtue by letting go of her
masculine walls, by going outside, or by letting a man in. She
can abandon such walls as long as she expresses her desire to
be at one with the Son of God or divine bridegroom dwelling
in heaven. Overtures to the Son of Man can replace the masculine ramparts covering her body. Given the right conditions,
signs of a woman leaving her house or letting a man inside her
dwelling can accommodate an ascetic scenario. A pious woman
can choose to step outside her door in the evening, provided that

the entrance is covered by clouds of heavenly glory (announcing
a bridegroom's visit) and that it is used for votive ceremonies
involving praying, weeping, healing, and cleansing (washing the
feet of the saints).[32] She can sprinkle the blood of a sweet offering
round about her household entrance in the evening, in accordance with plans of redemption by propitiation.[33] The imagery
points to a model woman or nation acting as a doorkeeper to
the house of God and guarding the portals of the Lord's dwelling
and city, the opposite of a woman dwelling in a house of ill fame
(Prov. 8.3, 34). The symbolism suggests a sleepless nation thanking the Lord for his wondrous creation and listening to voices
coming out of doors in heaven (Rev. 4.1).

While virtue may require that a woman step outside her
house, letting someone in may also be the right thing to do.
She may behave like Rhoda inviting the fugitive Peter into a
house of prayer. If so, she will emulate the faithful welcoming
Christ calling at their door, or God answering the prayers of
those knocking at his door, the door of bridal faith and heavenly hope.[34] A female doorway passage expressly devoted to
God opens on to a lady of wisdom cleared of all imperfections,
a body pure as the holy tabernacle. It opens on to God himself,
the gate of heaven: "I am the gate of the sheepfold. All others
who have come are thieves and brigands; but the sheep took
no notice of them. I am the gate. Anyone who enters through
me will be safe: he will go freely in and out and be sure of
finding pasture."[35] A woman welcoming God inside or outside
her walls is entering the gate of heaven.

The same logic applies to the man of God who lives in the
hope of wedding the "lady of wisdom" in accordance with the
law. He too must cover his body with signs of a desire to
possess the other gender, or sublime imageries thereof. This he
can do by entering the Mother Church, penetrating her temple,
wearing her robe, and occupying her Holy Land. Paradoxically,
faith is signified by breaking regulations of sexual distance,
transgressing rules that keep the two genders strictly apart.

Erotic metaphors are essential ingredients of ascetic language.
Signs of sexual activity are reproachable only when they lapse

A pair of sandals. Early Christian mosaic from a former
church in the village of Massuh, Jordan, reminding visitors
to take off their shoes and wash their feet. 6th C. (Copyright
Erich Lessing / Art Resource, NY S0147599 ART26790.
Mount Nebo Museum, Madaba, Jordan)

into licence, letting Eros be exploited without sublimation.
Thus removing one's shoes when going into a woman's house
or bed is not the same as taking them off when entering the
Lord's temple. While both shoe-removing gestures involve pleasurable dispossession, they differ in all other respects and are
mutually exclusive. We have seen that the erotic scene of a man
taking his shoes off as he enters a woman's house or bed is a
reminder that the feminine shoe-property occupied by a man's
feet is no substitute for the female enclosure he is stepping into.
The temple ritual (similar to that of a modern-day Christian

man taking off his hat in church) uses comparable imagery but with the opposite effect, which is to throw negative light on "real" erotic activity. That is, feet stepping into a holy place cast a shadow on limbs moving into a woman's house or a whorish temple. In the end, a woman's domestic interior is no substitute for the church temple prefiguring the heavenly abode. While both are erotic, the two scenarios of a man entering either a woman's house or the Mother Church are subject to the rank ordering of sign judgment: sublime love ranks above sexual and religious "fornication."

The scriptures are unequivocal about the imperfections of worldly pleasures. They are but metaphors for blessings in after-life. Symbols should not be confused with what they symbolize. Likewise with earthly gains: they should not be confused with the higher rewards they copy or imitate. Biblical displays of joys on earth never last. Even nuptial celebrations are but pale reflections of the wedding or love feast that comes at the end of time. Readers are reminded of Christ's reply to the Sadducees querying the fate of a woman who marries seven brothers who die one after the other. After her own death and presumed resurrection, "to which of those seven will she be wife, since she had been married to them all?" Christ responded that "at the resurrection men and women do not marry; no, they are like angels in heaven" (Matt. 22.23–32). Marriage on earth is a travesty or wishful simulation of the union of Christ and the model woman or the Church redeemed and elevated to the rank of a heavenly bride. Those trapped in purely worldly conjugations forego the surreal version of pleasures enjoyed in the hereafter.

Sacrificial imageries are transfigurations of Eros. Signs of the sublime and the surreal transport eroticism to a higher plane. The obverse is also true; erotic imageries carry ascetic value. This brings us to scenes of erotic activity proper, particularly those involving a lavish display of marks of gendered possession – man occupying and appropriating sites of feminine enclosure. Five observations are to be made regarding these erotic compositions, all of which pertain to "sacrifices" incurred through carnal activity.

1   We have seen that scenes of pleasure expressed through foot, shoe, and door symbolism often treat metaphorical possessions as pale reflections of the "real." This implies that metaphors may have to be relinquished or renounced if real pleasures are to be enjoyed (e.g., removing the shoe in a woman's bed).

2   "Real" possessions are in turn pale reflections of the sublime, surreal joys they resemble but can never replace. Deprivation remains unavoidable, even when carnal pleasures are secured.

3   Higher rewards are described in such metaphorical ways, using language borrowed from the blessings of life on earth, that they constantly recede into the distance. Blessings of the hereafter are all the more alluring as they remain deeply inaccessible.

4   Even when men indulge in gladness beyond morality, they do so with convoluted symbolism (e.g., removing the shoe) and related paraphernalia (e.g., wearing girdles of merry dancing). Sin is preferably expressed through metaphorical tokens of what men actually possess, in lieu of plain manifestations of lust, wealth, and power.

5   Last but not least, men can "possess" the other gender and cover themselves with signs of female gains on one condition: that they renounce their purely masculine selves. Male bodies must give up projecting attributes of their gender alone, to the exclusion of signs of feminine conquests and yearnings thereof. By dressing their bodies with traces and hopes of feminine winnings, male bodies divest themselves of claims of absolute masculinity.

All five effects converge on deficiencies and losses inherent in marks of pleasure. Signs of deprivation and deferral are constantly at work in expressions of Eros. As curious as this may seem, scenes of pleasures sliding into depravity require indices of deferment and self-renunciation, hints and metaphors of a shocking commerce but never the actual consummation. The attentions of Eros feed on elements borrowed from the language of restraint more typically associated with ascetic schemes, losses incurred through self-denial.

The first loss incurred through erotic sign activity consists in man foregoing "tokens" and "vouchers" of pleasure. Male figures are portrayed finding pleasure while taking off their shoes. The token footwear is relinquished and removed by virtue of its metaphorical value, because of the potential it has of being replaced by the "real thing" it stands for. It is taken off so as to mark men's feet entering into the womanly houses and wombs they seek to possess. The shoe then becomes a "pleasurable dispossession," a signifier foregone in exchange for the signified. In these scenes preference is given to losing the symbol (shoes) for erotic purposes in lieu of foregoing what it symbolizes (women) with ascetic intent.

Losses are inherent in expressions of desire, be they ascetic or erotic. But what happens to these losses once the man possesses the woman he longs for? Are they still operative? When enjoying the good life, deservedly or not, men are often portrayed as possessing everything they crave for. They may wear shoes not in lieu of something they lack but rather in addition to everything the shoe token stands for, other valuable possessions shoes normally "come with." Could we not conclude from this that a fully erotic scenario entails such gratification as to engender a desire to desire no more, a zero-degree longing that spells the end of metaphor?

Not really. After all, the attentions and expectations of desire are inexhaustible. Consider scenes that centre around shoes and doors, from a female perspective this time, scenes that tip the balance in favour of unrestrained pleasure and a clear satisfaction of gendered wants. Two radically different genres are explored below: the biblical, where displays of erotic connections show some restraint, and modern-day pornography, where visual imageries are given maximum licence. Both genres raise the same issue: given the free reign of Eros, why should there be losses of the feminine self, "self-denials" incurred through a profusion of masculine inscriptions covering and defining the female body?

Scriptural sites that connect licentious sex with shoes and doors are many. One such site consists in an impure woman opening her door to a lover stepping into her house and bed. When preserved, virtue covers the female body with masculine ramparts, signs of male possession she longs for. When a woman's virtue is sullied, male ramparts break down, yet this is only a prelude to a further invasion of masculinity within a woman's interior. Through this breach the woman simply replaces the "ramparts of her fortress" by the man and the lower limbs she welcomes in her house, bed, and womb.

The terrain an unfaithful woman gladly loses to the other gender can also be explored in scenes of a harlot leaving her house and running barefoot after her lover. A lady on the game chases the other gender like a male conqueror or hunter running after his female prey. In doing so, she abandons her own feminine attributions and corresponding shoe apparel and house enclosure. In the end, a woman no longer veiling, enclosing, or protecting her body and virtue with footwear, wall ramparts, and closed doors has one thing in common with a woman of virtue: both ladies must give up signs of their feminine isolation in order to communicate how desiring or desirable they wish to be.

The biblical scene of a woman decking her feet and body with adornments and jewellery is even more telling. Although erotic, the scene involves clothing in lieu of nudity. Instead of going naked and indulging in "beastly fornication," the woman is dressed up to the nines. How is it then that her behaviour is tantamount to lewdness? Why should this imagery betray a lack of modesty and moral worth? The answer lies in the masculine works of metal covering the feet of delicate women whose soles never touch the ground, women walking "with their heads held high ... tinkling the bangles on their feet."[36] Women of this kind do not divest themselves of clothing. Rather they cover themselves with signs of erotic possession. This they do through feet replete with masculine attributions wrought in the shape of precious metals. The cost to be paid is a renunciation of the purely feminine self. But given their intent to describe acts of lewdness,

why should the scriptures resort to so much symbolic convolu-
tion and circumspection? Why not a more straightforward
description of sexual activity? Why bother with a description
of the lewd woman's uplifted feet and tinkling shoe apparel?

Circumspection expressed through erotic footwear and uplifted
feet is symptomatic of biblical restraint. The scriptures cannot
indulge in outright descriptions of lechery lest too much attention should be granted to the unspeakable. But this is a weak
argument. The explanation would make sense were it not for
the fact that adorned-foot and high-heel imageries persist in
contexts that are less prudish. Modern-day pornography is a
case in point. Although showing bodies and sexual activities in
their full nudity, the attention that pornography draws to the
shoe and the foot verges on the obsessive. High-heel fetishism
thus begs the question: Why fetishize erotic activity if lewdness
is the message?

Some historical comments are in order. Although high-heel
shoes became fashionable in the mid-1950s, images of feet dis-
tancing themselves from the ground are not of recent origins. The
high heel made its appearance in eighteenth-century France,
with the heel under the arch of the foot. In England, narrow-
heeled shoes for women had become fashionable under the reign
of Queen Elizabeth 1. But the symbolism of female feet "separated from ground" is much older. The Book of Deuteronomy
(28.56) evokes "the tenderest and most fastidious woman
among you, so tender, so fastidious that she has never ventured
to set the sole of her foot to the ground." Given her infidelity
or lack of faith, the woman was condemned to suffer so much
misery and distress as to eat the flesh of her own husband and
of the children born from between her feet. Debasement and
lowly behaviour await women walking high on their feet.

Why mark the feminine gender with signs of aerial grace and
lightness? At first sight, the gendering process seems to be at
stake. Man is the heavy-footed or sure-footed gender firmly
planted on the ground or a rock (Ps. 40.2). Weighty he must
be if he is to trample the earth and put his footprint on the land
and property he possesses. Conversely, woman is the embodiment

of lightness, a delicate gender "supported" by the other and
elevated to the rank of the lofty, the angelic, the uplifted. This
gender difference is nowadays echoed by romantic scenes of a
man kissing and lifting a woman standing on tiptoe, the bridegroom carrying the bride in his arms, or the male dancer lifting
and throwing his female partner in the air.

High heels serve to demarcate the feminine from the masculine. They establish clear boundaries between the two genders.
The imagery nonetheless introduces planned confusion. A feminine gender vertically elongated can be suspected of being equal
to or "towering over" men who hold women in such high esteem.
In this manner the female gender is granted attributes otherwise
reserved to the male gender. The comment applies particularly
well to the stiletto-heeled shoe worn by women since the 1950s.
It too covers the feminine body with signs of a "higher position"
at odds with the normal rank ordering of Adam above Eve,
with man lording over the "weaker sex" (Gen. 3.16).

When examined closely, a high-heel shoe is slippery in that
it slides between contradictory gender attributions. On the one
hand, it abounds with feminine characteristics. The shoe is
fragile, delicate, and very slender. It covers the heel and the sole
but leaves a lot of flesh exposed: the instep, the arch, and
sometimes the toes. The stiletto footwear is thus an integral
piece of feminine accoutrement and a marker of *la différence*.
On the other hand, a lady wearing this shoe is not exactly a
prototype of the weaker sex. The shoe abounds with traces of
penetrating masculinity and offers women the prospects of a
"fashion with a vengeance" (Kaite 1995: 95).

All indications are that the stiletto or spike serves as a masculine writing of the body female. Meanings attached to this
style of shoe confirm the argument. For one thing the term
stiletto denotes weaponry. The term is the Italian diminutive of
stilo, a dagger, from Latin *stilus*, a pointed instrument, a style.
The dagger in question is small and has a thin, tapering blade.
The same word denotes a pointed tool used to pierce holes in
leather or cloth. It may be used as a verb, meaning to pierce,
or to kill by stabbing with a stiletto. Likewise, a spike is any

long, slender, pointed object, often made of metal. The word can signify a metal projection on the soles or heels of (typically male) shoes used for sports such as golf, baseball, or track, to prevent the sure-footed athlete from slipping. But a spike can also denote a long, heavy nail, projections along the top of an iron fence, or the unbranched horn or simple antler of a young deer. Lastly, to spike is to pierce or cut but also to add strength to a beverage, mixing it with alcohol. The man abusing such drinks may turn spiky and become quick-tempered.

Stilettos and spikes are phallic in that they point to extremities and things that are hard, long, sharp, piercing, and penetrating. In keeping with this male-styled imagery, a stiletto once signified a beard trimmed to a sharp point. Pointed associations of the stiletto are reinforced by toes and long hard nails partially protruding from the front of the sandal-like shoe. If toes are covered, the shoe itself will be pointed. (On this matter, it should be mentioned that toe points for men were fashionable in England during the thirteenth and fourteenth centuries and were so closely linked with status that the actual length of the shoes had to be regulated by law.)

To the extent that they emphasize the heel, stilettos and spikes point to the slang meaning of a heeled man: someone armed, especially with a gun. He may be not only well armed but also well heeled, propertied and wealthy. This man is the opposite of a shabby or slovenly person deprived of status, a man "down" or "out at heels." The money and phallic possessions evoked here tally with the support the long, stiff heel offers the foot, the extended upright leg, and the whole body. They also tally with signs of control and domination; witness imageries of men following close on the heels of others, lording over others falling under their heels, "spiking" and blocking their hopes of upward success.

In short, the stiletto shoe motif is eminently masculine and yet distinctively feminine, a stiff phallic extremity serving as a frail womblike enclosure. Given these ambivalent attributions, we can suspect the footwear of leading bodies and corpuses into active commerce between the two sexes. But to what end?

Why should so many convoluted connections move into opposite directions? Why use these sartorial schemes and schemings
of language to render the body erotic and attractive, seducing
one gender into paying attention to the other?

These questions can be answered by delving into the language
of seduction and metaphor, or the aesthetics of simulation and
the logic of desire. Secrets of eroticism lie in hints and promises
of the masculine and the feminine coming together. This implies
that each gender will cover itself up with signs of union of the
flesh, or hopes thereof. Each gender endeavours to cling to the
other so that in the end "they are no longer two, therefore, but
one body" (Matt. 19.5). Paradoxically, this language of desire
presupposes that each gender will surrender some of its attributions to the other gender. Each much consent to a partial
theft or dispossession of its gendered features. This is the only
way that man or woman can see the significant other possessed
with desire. Seduction lies in each gender's attraction to its
possession by the other, or the likeness thereof. The exchange
feigns a commerce or transaction that never comes to an end.
The final outcome is a perversion and travesty of coherent
sexual identities, an unsettling decomposition of the sexual
code invading all lawful pronouncements of *la différence*. While
exclusively assigned to the feminine domain, the stiletto is a
woman's simulation of the sexual encounter, a transgendered
prop used in the performance of desire.

The norm that says that only women can wear stiletto shoes
speaks not to the divide that separates man from woman: a
marker of difference acts rather as an invitation to commerce.
In its own self-contradictory "fashion," the gendered norm
speaks to hopes of boundaries being crossed. Normative regulations of the gender code are desirous and "trance-gendered"
from the start. Transgression is the law.

Women's stilettos do not betray the rigidity or fixity of gender
role attributions expressed through symbolic difference. They
have little to do with men's fear of castration, that is, their
inability to see in female bodies subjects that may exist without
signs of their quintessential masculinity. Nor are they fabrica-
tions of male subjects treating women as mere objects of desire,

reducing their bodies to men's shoe-like property, vulgar extensions and projections that men possess for their own satisfaction. More to the point, the woman's spiky accoutrement is an offering of desire. The fetish embodies a renunciation of the gendered self, an exercise in erotic self-denial that both simulates and stimulates a "small death" forever deferred. The shoe triggers hopes of desire. It is a tangible mirror image of the other gender's intangible desire.

A woman of small virtue wearing high-heel shoes and foot adornment wrought in metal does not surrender her body. Her body language delivers something even worse: an unrestrained expression of desire. The invitation or offer works on condition that her desire is what man desires. The imagery adds up to a woman actively pursuing man's desire of her desire. If this is true, then scenes that spell out the abominations and filthiness of "outright fornication" (Rev. 17.4) deserve little attention. Exposing the anatomy of sexual commerce would be inappropriate, especially in the biblical context. More importantly, words that lay out crude details of one gender possessing the other are limbically weak. They leave out signs of each gender being possessed with desire, signs that simulation of boundaries crossed alone can generate.

Transgression of gender boundaries is the rule. It is nonetheless a variable law. Eroticism and pornography may deploy comparable imageries, but the two approaches to sexuality differ considerably. A nude woman wearing stiletto heels may be a kindred spirit of the biblical lady walking high and tinkling the bangles on her feet, yet the two ladies are distant cousins at best. What separates them, apart from distance in time? Given the observations made above, we might say that the two differ not so much in levels of nudity as in revelations of sexual be-longings. The distinctive mark of pornography is that the body is fully surrendered to the yearnings of Eros, with no concession to other fields of desire. Sexual longings cover and hold sway over the entire body, including parts otherwise clothed or put aside for other semiotic actions. Breasts are no longer reserved for a mother's children; fingers for labour or for a wife's chosen man; lips and mouth for a pious woman's bridegroom and father dwelling in heaven (guarding her lips). Nor are the body's "private parts" hidden from sight, filthy areas to be concealed and washed away so that bodily needs are not confused with the higher functions of desire. With pornography, desire is dragooned into adopting a diet based on sex and sex alone.

Consider again the stiletto shoe used in a pornographic context. As Kaite observes, the high-heel shoe is a staple and prevalent sign in displays of soft and hard pornographic choreographies. "Regardless of the scenario, whether the model be in bed, on a couch, in a bathtub, or photographed in a 'nature' setting, the shoe is visible within the diegetic space" (Kaite 1995: 97). Intriguingly, "the shoe is in excess, often given to greater visibility than the breasts or genitals" (ibid.: 103). The shoe's phallic effect becomes all the more obvious as models bend their legs and point their heels towards the vaginal orifice, aiming (but not shooting) for penetration. Other signs of phallic projections include protruding nails, breasts, tongues, clitoris, and hair. When measured against the sexual act itself, anatomical and sartorial signs standing in for the phallus may seem somewhat superfluous. When measured against the logic of desire, however, they become essential. Without them coitus becomes rather lame.

The function of clothing is to conceal or dress up an erotic desire wanting to go despotic, beyond a disclosure of nudity and sheer anatomy. In pornography, models are stripped of the lingerie and clothing that normally keep erotic sign activity in check, placing limits on signs of longings to be possessed by/of the other gender. Pornography lifts the veil not on the body we possess but rather on Eros possessing our body. Therein lies the insult to western morals: not in the unbridled satisfaction of carnal pleasure but rather in the shocking reign of erotic desire.

With pornography, older signs of Eros (e.g., uplifted feet) are sexually focused, hence pushed into saying less than what they were previously allowed to say. X-rated imageries block off the many circuits and connections deployed in older scripts. Eros gone strictly sexual is made to reign over all protrusions and orifices of the body. Anal eroticism is particularly instructive in this respect. As Kaite (ibid.: 97, 100) notes, female models displayed in pornographic magazines are often shown "mooning" the camera, with their rear end inviting ocular intrusion, sharp fingernails spreading the buttocks, the leg and knee bent towards the back with the stiletto shoe nearing the anus. Why these anal adaptations? Two effects come to mind. One is to debunk all notions of biological functionality – the idea that some parts of the body serve functions other than sexual pleasure. Another is to dispel signs of immorality vested in sights of filth and the shameful parts. The implication is that parts of the body connected with shit, dirt, and soil (hence the anus and the foot) may be granted erotic connotations. This is tantamount to saying that filth and things dirty can be pleasurable, the pornographic premise par excellence. Erotic filth needs no washing and ablutionary foreplay, after all.

We know that colloquial expressions evoking excrement and the anus are overwhelmingly pejorative. They are as negative as the "heel," which can mean an unpleasant, despicable, or contemptible man. People viewed as "assholes" and "pieces of shit" are rarely envied. No one wants to be "in the shit," be a "shit-disturber," or see the "shit hit the fan." People don't want "shitty things" to happen, at least not to them. Likewise, it is unusual for someone to brag about the fact that he or she is "living in a shit-hole." This said, individuals may still hope to "relieve themselves" of the burden of a "shitty life." They can aspire to become "well-heeled" and "filthy" rich. In other words, they can let themselves be tempted by the rule of desire, pushing the reign of pleasure beyond parts of the natural body reserved for "clean sex." In pornography one "shameful activity" absorbs all others.

To recap, signs are malleable. They lend themselves to variable connections that are responsive to broader reticular contexts. In the Bible, lower limbs of the body and related imageries such as footwear can adapt to multiple scenarios ranging from the joyful to the sinful, the tribulational, and the sacrificial. These

Chair, Allen Jones, 1969 (Copyright Tate Gallery, London / Art Resource, NY S0089469 ART83901. Tate Gallery, London, Great Britain)

signs are pliable in yet another sense: marks of possession or dispossession (wearing or removing the shoe, gaining or losing the firmness of leg and foot, etc.) can be tied to them without altering their overall effect. Whether something is gained or lost is relatively unimportant. What matters is the tactic that is being served, be it through measures of positive or negative "disposal" (having access to, getting rid of). One such tactic is of the diacritic kind. It consists in using cognate signs to announce or stand for one another but without appearing at the same time (e.g., removing the shoe if entering a woman's house). The other tactic is of the syncretic kind: letting cognate signs either come or go at the same time, be they pleasurable (joyful, sinful) or not (tribulational, sacrificial; e.g., wearing the shoe when taking a brother's wife in marriage). Both tactics are governed by the order of desire, which operates at two levels: diacritic and syncretic strategies can be used to evoke things deemed to be limbically attractive, but they can also signify things considered to be morally wholesome.

One puzzling implication of this argument is that desirable things can be relinquished "with pleasure." Also they can be possessed "with pain." The imageries we have explored confirm this tendency for ascetic and erotic imageries to be deeply entangled. They intersect for two major reasons. First, effects of self-denial can be obtained by covering the virtuous body with erotic "be-longings" (e.g., shoes worn by missionaries in exile), tokens of what the body relinquishes and continues to long for (e.g., the Promised Land). Indices of erotic pleasure thus make their way into scenes of ascetic conduct; transgression becomes a mere in-fraction of the law. Second, effects of self-gratification may require that signs of one's gender be surrendered to the other – letting the other cover itself with hints of masculine and feminine intercourse, hence invitations to erotic commerce. Traces of sacrifice verging on devotion and devoutness slip into circuits of licence and corruption.

Through the simulations of metaphor, pale reflections of the surreal cover the body with attractions and approximations that prefigure and promise blessings that are never fully delivered. The real thing never comes. Nor is "real" sexual identity ever communicated in its naked truth, stripped of its desirous possessions and be-longings. Identities differ essentially by virtue of what they do not have and what they yearn for, "be-longings" written all over their "trance-gendered" selves. Despite its claims to full anatomical truth, pornography itself does not escape this compulsion to simulate the desire of the other's desire. While Pornos may put desire on a strict diet (sex as the main dish and nothing else), it is not empowered to alter the fetishistic gastronomy of Eros.

From Earrings to Body Piercing

Erotic displays of female desire point to the male figure taking pleasure in a masculinized woman simulating the possession she longs for (and vice versa). The language of seduction betrays a woman seeking out a man longing to conquer her desire through a partial simulation of the union he and she long for. Her yearning to be possessed becomes the man's valuable possession, a confirmation of her desire for his desire. Correlatively, the male figure takes delight in giving up and surrendering *fragments* of masculinity to the female subject (high heels, long nails, precious family jewels, etc.). The erotic act thus demands that sacrifices of gender be made from the start.

The language of desire is a prelude filled with signs of sexual deferment and self-renunciation. Eros is a worldly "play of passion" in which a code that distinguishes one gender from another (keeping the two minimally apart) is combined with signs of commerce bordering on a loss of sexual identity. Signs of erotic "self-denial" contravene the code of sexual distance. Gender symbolism is rife with the offerings of a small death, gestures of seduction at variance with the Freudian fear of castration. Sexuality is more often in collusion with morality than not.

These findings have a direct bearing on central debates in the field of semiotic theory. Before we explore these implications, however, more should be said about historical variations and shifts in the language of body and desire. As already pointed out, biblical formulations of gender transactions cannot be simply lumped together with modern expressions of Eros. Nor should present-day conventions of "straight" eroticism be confused with rules of the pornographic genre. Erotic assemblages are highly malleable and produce effects that vary considerably and that speak to broader polemics of social and cultural history.

Take masculine possessions covering the female body. They can take many shapes. They include not only footwear but also works of metal and jewellery adorning different parts of the body. Jewellery can also pierce the flesh. Earlobes pierced or clipped with metal have gone through recent changes reflecting current statements about the body, but have a long history taking us back to biblical themes linked to shoes, houses, and doors. As with shoes, pierced lobes used to be directly tied to biblical matters of property and possession. This is confirmed in Deuteronomy's slavery laws modelled on ties between God and his faithful servants: "If your fellow Hebrew, man or woman, is sold to you, he can serve you for six years. In the seventh year you must set him free, and in setting him free you must not let him go empty-handed. You must make him a generous provision from your flock, your threshing floor, your winepress; as Yahweh your God has blessed you, so you must give to him. Remember that you were a slave in the land of Egypt and that Yahweh your God redeemed you; that is why I lay this charge on you today. But if he says to you, 'I do not want to leave you,' if he loves you and your household and is happy with you, you are to *take an awl and drive it through his ear into the door* and he shall be your servant for all time. You are to do the same for your maidservant" (Deut. 15.12–17; see also Exod. 21.1–11). A person fastened with iron to a door belongs to the owner of the house.

Earrings signify possession. Under what conditions should this possession be understood erotically, as marks of desire rather than effective property? Consider earrings attached to the female gender in biblical texts. Earrings worn by women bring two things together. On the one hand, the jewellery marks an earlobe or piece of boneless flesh that serves no apparent physiological function, none culturally recognized, at least. The lobe is thus in an ideal position to stand for pleasures of the flesh enjoyed for their own sake, without biological purpose. This is the soft flesh that men succumb to when seduced by sins of the "weaker sex." On the other hand, precious stones and metal works fastened to the earlobe are hard and penetrating. They may serve as indicators of a man's material wealth and may be given to women by men as tokens of their love. The metal earring and pierced earlobe imagery thus adds up to a piece of feminine flesh decked with an array of masculine possessions and be-longings. The adornment is yet another simulation of gender intercourse appealing to the erotic imagination.

Scriptural versions of this imagery are worth mentioning. Biblical evocations of the blessings of wealth and beauty associated with earrings are far reaching. When pleased by her conduct, God rewards the model woman by spreading his skirt over her. He covers her nakedness, washes her with water, and dresses her in fine linen. He also decks her with precious earrings, bracelets, necklaces, amulets, metal charms, and a beautiful diadem (Ezek. 16.8ff.). The woman's flesh is thus covered and mounted with the metallic wealth and strength of a man of great heavenly "charm," so to speak. The attractiveness of this woman, however, does not come from the intrinsic worth of her glowing apparel or precious assets obtained from the opposite gender. To be sure, she is earmarked for the good life and the enjoyment of possessions granted by God. What is more important, however, is that she be marked for the possessive love of the Man of God. This can be done by letting the woman put on a semblance of the Holy Communion and heavenly wedding she yearns for. Beauty does not stem from the woman partaking of her master's valuable belongings. Earrings and jewels received from her Lord serve rather as tokens of her yearning to come into God's possession. Her desire for God's love is what her Lord desires above all. The woman's desire for her Lord's desire is made manifest through an erotic simulation of the wishful disposition itself, of her longing to be joined with the Son of Man in wedlock.

God rewards a virtuous lady by decking her ears with jewellery. But the Lord will recognize the deserving woman through

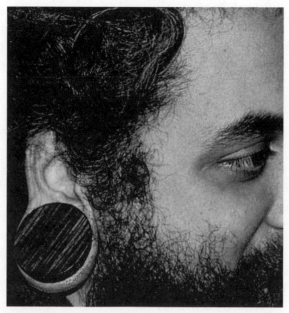

Ear plug by Blake Hipson, Classic Body Piercing 11, Ottawa, Ontario

her show of virtue, which implies her refusal to wear fine jewellery. If never surrendered and offered to the service of God (Num. 31.50), precious earrings are undeserved and point to a life of sin. A woman of virtue should therefore dissociate herself from a woman of the street offering bell sounds, talismanic "earrings" ("incantation" in Hebrew) and other amulet charms to gods of the flesh (Gen. 35.4). The damsel must show her virtue through modesty lest she should move God to anger, be stripped of her fine apparel, and fall from grace. Valuable earrings can be lastingly possessed only by those who divest themselves of such goods.

Scriptural usages of the earring motif point to a twofold commerce between the masculine and the feminine and between gestures of possession and dispossession. Displays of the male gender surrendering some of its masculine imagery, covering the feminine flesh with hints of intercourse, feed on plays of desire and seduction. But scriptural usages of the ear imagery

do not tell the whole story. While the story of clipped and
pierced ears goes back a very long time, variations in time and
space should not be underestimated. One such variation can be
found in recent changes affecting western body piercing practices and the current pluralization of sexual lifestyles. We know
that in "straight" body language, only the female subject clips
or pierces the earlobe, producing meanings and implications
compatible with the interpretive comments presented above. In
recent decades, however, simple gender divisions have been subverted and parts of the body earmarked for erotic suggestions
have been multiplied. Men will now wear earrings, some in
their right ear to signify that they are gay. Also both men and
women, especially the young, will pierce their eyebrows, noses,
lips, tongues, nipples, navels, and genitals with rings and jewellery. Why this current trend towards multiple piercing and
sexual indifferentiation?

Wojcik (1995: 35–6) suggests that body piercing and mutilation is a modern revival of primitive aesthetics, a neo-tribal
art form socially disturbing by design. Body piercing challenges
the current regime because of the pain, masochism, and self-
destruction it entails. The art is also an expression of estrangement from mainstream society, a quest for identity without
conformity. Finally, self-mutilation betrays a desire to gain aesthetic and erotic pleasure through control and transformation
of the body. In a similar spirit, Vale and Juno (1989: 4) see this
form of control as an admission of the subject's powerlessness
to change the body social and world in its current state.

These explanations are insightful in that they emphasize the
symbolic expression of debates regarding dominant forms of
body language and social life. They are nonetheless problematical in two respects. For one thing they neglect the fact that
bodily mutilation is not absent in earlier fashions, albeit in
milder forms. More importantly, the precise ways in which conventions are subverted and the extent to which particular body
piercing practices codify this challenge are insufficiently explored
(beyond their mere description). Phenomena of estrangement,
powerlessness, and body control through self-mutilation are

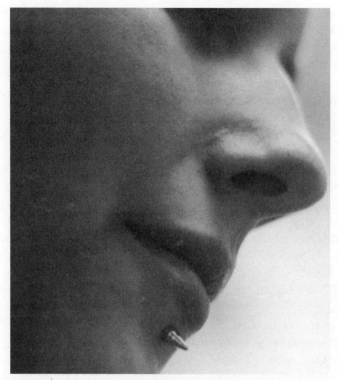

Lip piercing by Ian Wright, Classic Body Piercing II, Ottawa,
Ontario

simply too general and widespread to account for the main features and idiosyncrasies of current body piercing practices.
Questions such as why pierce the lip or belly button instead of
the traditional earlobe must be answered lest the analysis should
fall into generalities of social psychology.

Briefly, I suggest that the current subversion of body aesthetics through self-mutilation is to be understood against the background of two related processes: the division of sexual labour,
and the sexual division of labour. The division of sexual labour
concerns relations between genders and parts of the body. It
informs not only body piercing in general but also particular
variations of this art form. In the "straight" genre, a distinction

is made between body parts that may act as erotic markers and those that may not. Paradoxically, parts of the body that are assigned erotic functions through adornment do not include the sexual organs. Nor are they limited to bodily orifices (mouths, eyes) and extremities (fingernails, feet) standing in or substituted for "private parts" covered with clothing. There are other parts that receive decorative attention and yet have no function to serve other than erotic. One suspects that it is precisely because of their uselessness that they can be singled out for the labour of seduction. Pleasures of the flesh and commitments of the heart must show physiological disinterest. The language of seduction is at odds with mere biology or signs thereof.

The ring finger on the left hand confirms our suspicions. It stands as the least useful and dexterous of our ten hand extremities. Accordingly, it becomes our most amorous digit. Earlobes are equally revealing in this regard. They are fragments of "pure flesh" serving no apparent function. Given their superfluousness, they are apt to be earmarked and pierced with signs of masculine possession: that is, earrings made of precious stones and jewellery fastened with rings, screws, and clips. In the "straight" genre, female orifices such as eyes and lips can be highlighted with cosmetics. Likewise, female extremities can catch the male eye with nail polish and high heels. Until recently, however, a woman's eyes, lips, and limbs were left unpierced because of their functional services. The organs and zones in question were employed in the works of sensory perception, physical labour, language, and communication. Body parts dedicated to "useful" activities were not eligible for signs of full erotic possession.

The 1950s brought about meaningful deviations from premises of the "straight" genre. For one thing, the introduction of stiletto heels eroticized the female body beyond earlier conventions. The move coincided with two broader trends. One involved shifts in the sexual division of labour, with a greater participation of women in the labour force, reducing the confinement of their bodies to reproductive chores and domestic activity. Another trend consisted in a new ratio of production

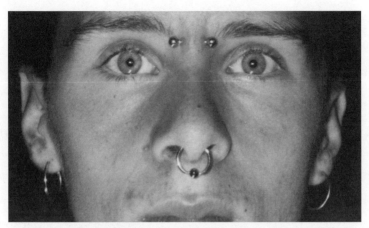

Multiple piercing by Ian Wright, Classic Body Piercing II, Ottawa, Ontario

to consumption activity. Dreams of the Affluent Society promoted leisure and the enjoyment of goods and services of all kinds, women's sex appeal included. With the hippie movement and cultural upheavals of the 1960s, these trends were extended to a more open exploration of alternative sexual lifestyles and gender role attributions. Accordingly, body piercing aesthetics were used to mark the male body's fuller entry into the language of carnal activity, be it through free sex or gay "coming out" events. Gays signified this by perforating either the right ear (unlike women) or both ears (like women). Heterosexual men started piercing their weaker ear lobe (the left) with signs of the masculine taking possession of their female side. The implication of a masculine-left-earring inscription is that man is willing to let his subsidiary femininity appeal to and be possessed by woman's subsidiary masculinity. A man's equal, the liberated woman, can attract and be attracted by a woman's equal, the liberated man.

"Neo-tribal" aesthetics of the 1980s and the 1990s represent an even more radical shift. Severe recession and unemployment crises resulted in the demise of the welfare state and the Affluent Society. For some these developments have brought about a

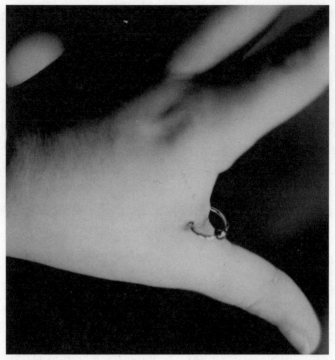

Hand piercing by Ian Wright, Classic Body Piercing 11, Ottawa, Ontario

partial reversion to conservative morality and older work ethics. For others, especially the young, disillusionments of late
modernity have meant a rejection of basic codifications governing all divisions of labour, productive and reproductive alike.
Dualistic regulations of bodily transactions no longer hold.
Body piercing is part of a broader rebellion against hegemonic
inscriptions that separate the masculine from the feminine, consumption from production, instrumentality from expressivity,
the private from the public, dirt from cleanness, and pain from
pleasure. In the straight genre, only female parts of the body
deemed to be non-private and to serve no instrumental or productive function could be surrendered to the ephemeral pleasures of seduction. The primitive-modern body is now carved up
differently, with "shocking" indifference to these conventional

limitations and binary subtexts. All bodily areas and functions can be marked out for non-stop erotic activity. In the same breath, all activities or functions other than erotic are obliterated: the body becomes a pornographic corpus.

With body piercing, metal perforating all parts of men's and women's flesh is used to minimize or attenuate sexual difference. Flesh is no longer gendered. In the same rebellious spirit, sexual pleasure invading all nooks and crannies of the body is maximized. Perforations of the body no longer recognize distinctions between pleasure and suffering, let alone between sexual organs and their erogenous substitutes or simulacra. Pain and genitalia can be penetrated by sex, just like everything else. Marks of seduction were previously at odds with the fixed aspects of biology and were restricted to areas of public gaze; they are now extended to all the "private parts" of an erotic imagination aspiring to ubiquity and irreversibility (body piercing and mutilation are for life). Correlatively, body piercing does away with former divisions between useless, useful, and shameful parts of the body. Erogenous possessions are applied to protrusions and orifices (lip, tongue, eye, nose, anus) formerly reserved for productive labour, communicative activity, and bodily functions. Received lessons of decency and indecency are scorned, and older deliberations of filth and clean sex are written off. As with stilettos aimed at the anus, images of rings perforating organs, orifices, and protrusions make a mockery of conventional differences between male and female, mutilation and reproduction, pain and pleasure, labour and lust, reality and simulation. Postmodern bodies are shot through with signs of sexual tyranny.

All sign actions are polymorphous and transgressive in some fashion, yet only some involve far-reaching reassemblages. Prefrontal judgments are tinged with limbic affects, and emotive forces with normative constraints. Attentional mixtures nonetheless vary greatly in intensity and longevity. Some arrangements last a few centuries, while others are condemned to keep up with the fashions. With body piercing, the appeal and life expectancy of the modern era's "organic constitution" seem to have reached their limit.

PHILOSOPHICAL CONSIDERATIONS

The Foldings of Metaphor

Sign reticles (referred to as *Sr* below) are highly malleable. When seen from a sagittal view, some sign actions converge through RH syncretism, whereas others generate effects of divergence through LH diacritic processing. On the axial plane, sign connections produce variable attentions, depending on the event at hand. Some communications are fully attended, resulting in what could be called the soma expressions of semiosis. Others are given a secondary role obtained through autonomic depolarization, activation without full noticing. These ancillary transmissions constitute the background impressions and implications of "signaptic" events. All remaining connections are either polarized or hyperpolarized, unattended through active neglect or forceful inhibition, two forms of "sign capacitance" needed to ward off irrelevant or menacing connections.

When applied to linkages between cortical and subcortical functions, the attentional economy accounts for differences between reticles devoted chiefly to prefrontal normativity (moral, instrumental) and those assigned primarily to limbic affects driven by the pleasure principle (and the avoidance of pain). Sign potentials, however, cannot be pigeonholed into distinct functions and discrete areas of the brain. Sign events are always performed through the coordination of the numerous functions, centres, and levels of semiotic activity. Thus in *3-D Mind 1* we saw how sign actions following a syncretic pattern – for example, proper naming – contain elements of diacritic processing. Christ may resemble and be called a lamb

The Vision of the Lamb (surrounded by four evangelists and twelve of the
twenty-four elders, Apoc. 4:6–5:6–8). Commentary on the Apocalypse by
Beatus de Liebana, Spain (Leon), c. 950 (Copyright the Pierpont Morgan
Library/ Art Resource, NY S0114662 ART108986, CE. M. 644, f.87)

on the condition that Christ *is not* a lamb. Signs of convergence
require divergent supplementation in order to make sense. Likewise with diacritic events: they can take place provided the
brain apprehends a common terrain within which oppositions
or deviations can play themselves out (the Lamb slain and the
Beast are animal metaphors, sign-manifestations of disembodied
spirits dwelling in non-visible space).

In this book, I have extended the same multimodal reasoning to *communications performed along axial lines*. Although compelled to specialize, all sign reticles combine normative and emotive aspects. Semiotic effects ruled exclusively by limbic *or* prefrontal functions are the exception. Sign impulses that take a predominantly normative orientation are shaped by emotive forces, supportive and transgressive affects vested in the moral teachings at hand. The argument holds true in the opposite direction as well. Sr activities exhibiting high levels of affect contain adjustments and concessions to normative concerns, allowances that are essential to the deferrals and protracted invitations of desire.

In the preceding analyses, two "disposal" modalities were explored in regard to the interfacing of norm and affect. Signs of grasping or seizing attributed to the LH were compared with those of removing and ceding reflecting a RH disposition. The comparison between scenes of possession and dispossession was grounded in illustrative analyses involving fetishistic foot and body piercing imageries. Our interpretations of this material brought the malleability of nervous sign processing once again to the fore. Although possession is usually linked to self-gratification, and dispossession to self-discipline, both modalities are conducive to utterances of ascetic and erotic disposition. More importantly, two factors other than the "disposal" modalities were shown to determine the actual orientation (joyful, sinful, tribulational, sacrificial) of sign affects and norms. One factor resides in the explicit configuration of Sr connections – whether the script is removing the shoes from the feet of a man in exile or from those of a man stepping into a house. This factor does not suffice in determining affect and meaning. The other factor must also be considered and concerns the two-sided nature of metaphor.

One side of metaphor echoes the RH convergence rule; this is the side that allows things that come together to flock like birds of the same feather. The shoe metaphor used in this fashion summons attention to what it resembles, "belongs to," "comes with," and "is part of" – for instance, a biblical shoe that comes with property in land. Another example is a shoe that resembles a uterine house and turns into a shoe-house for honeymooners. The other side of metaphor is quite different. It entails an act of simulation, not to be confused with signs of similitude. This is the effect we obtain when the shoe metaphor gives us notice of what is absent, something that is feared or longed for (e.g., men in exile wearing shoes while travelling to the Promised Land). By definition, to simulate is to produce a *superficial* resemblance, to assume the appearance of something *falsely*, or to imitate something through a *different* medium. Unlike statements of sheer resemblance, the act of simulation draws our attention to the (LH) distance or divergence that lies between the sign doing the imitation (e.g., aircraft model) and the sign it imitates (e.g., the full-size aircraft). It is the distal aspects of metaphor that receive our maximum attention.

On the cognitive plane, a metaphor is primarily syncretic; it emphasizes proximal resemblance. On the normative and emotive plane, however, the amplitude of this syncretic effect can be reduced considerably, towards a diacritic slant that triggers desire by means of simulation. Metaphor is thus in a position to perform complex connections on both planes: the axial interfacing of norm and affect, and the sagittal processing of similarities and differences. The rule of metaphor thus operates at "*the junction* – not the unity – of ... a complex set of both horizontal connections and of vertical integrations reacting on one another, as is shown by cerebral 'maps'" (Deleuze and Guattari 1994: 208).

Logic and the order of desire are deeply intertwined. The argument further theorized in what follows is that no fruitful distinction can be drawn between the logical and what may be called "the desirable," as defined through signs of pleasure or acts of judgment and morality. Cognitively oriented studies of sign systems that neglect the emotive aspects of their subject matter are incomplete. By reducing sign action to semantic codes and cognitive grids, such endeavours project a premise of the scientific enterprise on to their object of study, namely, a desire to desire no more, wishing the intensities of affect out of the intellect (Lyotard 1993: 14–15).

The cognitive weavings of similarities and differences are shot through with normative and emotive threads. Logic, desire, morality, and rationality are inseparable. This point is often missed in theories pertaining to the logic of metaphor. The multiple facets of sign activity are largely ignored in what is known as the *substitution theory of metaphor*, as in Genette (Ricœur 1977: 129).

Briefly, this theory of Aristotelian origins holds that a metaphor involves the borrowing of a name or noun to name something figuratively, in a "so to speak" manner. This deviation from ordinary speech and the language of literal denotation occurs for one of two reasons. A metaphor may be used because of a semantic lacuna, forcing speakers to extend one term to another meaning (thing or idea) for which there is no proper denomination. Although borrowed, the secondary meaning assigned to a term may be so clearly marked by the context ("legs" to support a chair) that it leaves no room for alternative interpretation. If so, the term ends up acting as a "proper name" that may have origins pointing to another semantic field (anatomical, physical) but without these origins being actively grasped or noticed. As with the words "grasp," "conceive," "figure" or even "metaphor," some tropes are eroded and burdened with age, figures of speech that become habitual, standardized, and fixed. Thus we speak of the Milky Way without having dairy products in view. Commonplace expressions that have undergone advanced lexicalization act like literal meanings, moving away from the metaphors that originally inspired them, closer to catachreses based on misunderstandings or losses of etymological memory (ibid.: 81, 100, 110, 121–2).

Alternatively, a metaphor may be substituted for a literal expression or meaning signified *in absentia*. This occurs when a denotative term or exhaustive paraphrase is available but foregone for reasons of stylistic choice. The implication is that the foregone meaning can be reintroduced *ex post facto*, for purposes of literal restoration and explication. Thus the "flame" trope ("an old flame") can be translated back into the "love" it stands for, thereby "correcting" the metaphorical deviation from signs of the "proper" code (ibid.: 17–19, 46). Unlike Aristotle, proponents of substitution theory claim that metaphors are rhetorical and aesthetic adornments that play no predicative or discursive function. They add no information to what is being indirectly named. In its own way, substitution theory is a derivation of cognitivism and nominalism. Metaphors are "so-to-speak" usages of denotative language, hence "mere figures of speech" (ibid.: 33–4, 46, 88–9, 139).

Substitution theory centres on the relationship between metaphors and the literal denotations or descriptive paraphrases they replace. The assumption is that metaphors are words or expressions twice removed from the extralinguistic concepts, objects, or subjects they signify. They act as signs that deviate from their proper meanings in order to signify something not falling under their own denotative jurisdiction. Metaphors point to words that point to things lying outside the realm of language. The whole reasoning revolves around what Richards calls the "proper name superstition" and figurative deviations thereof (ibid.: 77–8).

The alternative view defended in this book is that metaphors are predicative relations between fields of meaning, hence assemblages in their own right. The argument is in keeping with theories of signs networking in language. As Deleuze and Parnet (1987: 117) suggest, there are no metaphors, only combinations. To use a biblical example, the Lamb slain metaphor applied to Christ in the Book of Revelation presupposes not a deviational substitution of an animal motif for the proper name (Jesus Christ) designating the Son of God; more to the point, the imagery rests upon the intersection or superimposition of relations taken from two differential domains: that is, differences between animal species and differences between spirits. Christ is to the Antichrist what a lamb is to "beastly" animals such as scorpions, locusts, and jackals moving in the train of the Antichrist (Rev. 9). Metaphors draw parallels between fields of meaning, in this case the zoological and the spiritual. As in totemism (seen from a Lévi-Straussian perspective), metaphors are correlations of opposites based on similarities between differences.

A metaphor interfolds one sign plane with another; one plane folds around a fold of the next plane. Through interfolding, a metaphor activates not only the sign actions appearing on the surface script (Christ and Antichrist, lamb and scorpion) but also the implications mapped on to each fold (e.g., the similarities and differences between lambs and scorpions). Conventional associations attached to signs within their respective domains are intimated and mobilized through metaphor. Thus while lambs are domestic animals meek and valuable, scorpions and jackals are wild, aggressive, and deadly. By "interfold" extension, Revelation applies the same contrast to the antagonism between Christ the sacrificial figure and Antichrist the evil destroyer. A simple predication results: Christ is to Antichrist what a sacrificial animal is to a murderous beast.

According to Aristotle, metaphors are based on proportional analogies: a is to b what c is to d. Thus, in English parlance, wings are to birds what fins are to fish; the "evening of life" implies that old age is to life what evening is to day. But metaphors can also be based on substitutions from genus to species, species to genus, or genus to genus. This means that metaphorical effects can be obtained by correlating two sign levels within a common domain, substituting general differences for particular ones (or vice versa). In the English rendering of biblical imageries, the term "beast" refers to a wide range of four-legged animals. Yet the generic term, the Beast, can be used to stand for a particular animal species, namely, the unholy as opposed to the holy. Thus we may say that jackals are more "beastly" than lambs. Thus jackals (A1) are to lambs (A2) what beasts (A) are to spirits (B); A1 is to A2 what A is to B. These "intrafold" metaphors involve the mapping of species on to genus differences (or vice versa).

Tyconius, an advocate of the fourth-century Donatist movement opposed to Christian Church accommodations with Rome, played an important role in studies of "intrafold" metaphors and exerted a major influence on western medieval exegeses via the works of Augustine. He applied a historical understanding of apocalyptic prophecies to the North African millenarian battle against the state church and other worldly embodiments of the Beast. Despite his commitment to millenarianism, however, Tyconius denied that the precise timing of the End and the literal identity of the redeemed could be known; he recognized a limit to correlations that can be read into the language of metaphor. This agnostic attitude towards the timing of God's interventions in history was derived from a larger hermeneutic agenda developed in his *Liber Regularum*.

In this book, Tyconius identifies seven basic rules of sign substitutions that govern scriptural meanings and reflect the anagogical mysteries of biblical knowledge. Three of his substitution rules are relevant to our discussion. They can be summarized as follows:

1   *De Domino et corpore eius:* When the Bible speaks of the Lord, it may mean his body, the Church.
2   *De diabolo et euis corpore:* Passages that speak of Satan may intend his body, hence his followers.
3   *De specie et genere:* scriptural references to events (*specie*) may serve to express general truths (*genere*).

In the scriptures, terms are often transposed from one level or generality to another, thereby serving metaphorical purposes and requiring restoration through proper exegetic interpretation.

The taxonomy of figurative substitutions proposed by Aristotle and Tyconius overlooks the logic shared by proportional analogies and genus-species substitutions. As already shown, both bring different folds together. The former does so through similarities established between fields of meaning, and the latter through parallels between levels of significations.

All in all, a metaphor is not simply a word standing for a non-figurative term, a sign misused with heuristic intent. A metaphor acts rather as a deviation from the internal strictures of lexicologies, loosening one field or level of signification and projecting it on to another. Figurative connections stray away

from strict taxonomy attributions and immediate lexical associations (Ricœur 1977: 4). They do so for purposes that are
essentially connective and compositional – hooking up planes
of differences. The question, however, is whether these connections can make sense without responding to context through
the referential functions of language. Through metaphor, one
might argue, sign utterances are tied to the predicative attributions of full sentences and the contextual circumstances of
living speech. These issues of semantic and hermeneutic context
are discussed in the next reticle.

Beyond Semiotics, Semantics, and Hermeneutics

Substitution theory views metaphor as a denotation by proxy. As a result, it neglects the compositional assemblages and predicative aspects of metaphor. The act of calling Christ the Lamb slain may not be marked with signs of full propositional grammar (subject, verb, complement, etc.). The metaphor nonetheless entails an act of comparative predication: Christ is to his enemies what a Lamb slain is to jackals and scorpions. If spelled out, the comparison implies that the Saviour is meek as opposed to cruel, someone destined to be killed as opposed to being a killer, and so on. Metaphors are not connections used for mere decorative purposes. Rather, they generate attributive compositions, assemblages that are brief but nonetheless informative. One field, the metaphorical, adds something to another field, the metaphorized.

This predicative aspect built into the rule of metaphor suggests that all figurative acts generate semantic effects. As argued by Ricœur (1977: 65, 104, 123), a semiotic approach *à la* Lévi-Strauss or de Saussure is limited and calls for the added value of a semantic analysis *à la* Benveniste. Tropes cannot be reduced to the semiotic order of similarities and differences, logical combinations, and transformations of the code. They must also be understood within broader contexts that animate them, hence the larger sentences, discursive frames, and narrative stories that support them. In the Lamb slain example, a metaphorical expression will be understood when situated in its proper context, which consists of an apocalyptic narrative pitting Christ against the Antichrist and the beastly creatures in its train.

Ricœur reminds us that there is more to language than sign taxonomies providing codified references to things, ideas, and events. Signs interact among themselves, in the broader context of full sentences and related attributions and judgments of language. The reminder is in keeping with Wittgenstein's critique of the naming game consisting of dictionary entries, authorized definitions, genus and species descriptions, word-to-word translations, and concept characterizations. Semantic operations must also be taken into account. They have a direct impact on word references and how utterances are used in particular discursive contexts. Far from being governed by lexicology and code, meaning results from the referential interanimations of word and sentence, hence semiotics and semantics. Individual words may refer to objects, but they are also responsive to sentences that are tied to "states of affairs" conveyed through discourse. Sentences determine what words mean and referents that are contextually acceptable (ibid.: 128–30).

Ricœur also insists on the ontological import of figurative language and sign events. Language is never self-sufficient. All acts of signification are subject to the Husserlian principle of self-transcendence: they contain the imprint of intentionality, intersubjectivity, and mediation between humans and the world they live in. Room must therefore be made for the "higher" hermeneutical connections that harness metaphors and text to the nonverbal context in which they emerge and evolve. Accordingly, the interpretive practice must factor in the extralinguistic world and history humans dwell in, which includes the psychological and sociological circumstances of real life situations and events (ibid.: 92–3, 111, 123, 134).

The lexical interconnections and the internal organization of signs make no sense when detached from experiences that are external to language and to which the sign process constantly responds. To use the same example, the Book of Revelation must be understood with reference to the first-century battle of Christianity against Rome. The sense and meaning of John's

Apocalypse are grounded in first-degree references consisting of literal denominations. As in all products of language, the role of these referential identifications is to name singularities through demonstratives, pronouns, proper nouns, and descriptions of the particular and the concrete. The singularities thus named also include concrete actions, qualities, and class properties. All of these first-degree references ensure that products of language will effectively "stick" and adhere to reality (ibid.: 70–6). The denominations produced at this level are reliable and replicable not because words have meaning in themselves but rather because of the constancy vested in their referential context. Sense and meaning are not for all that reducible to literal communications. Signification also thrives on second-degree references generated through metaphorical language, signs that redescribe the world we live in, in accordance with rules of symbolling (ibid.: 5).

In short, sign interpretation as characterized by Ricœur goes beyond semiotics and delves into the higher-order linkages of semantics and hermeneutics in order to account for the interplay of word and experience. Consideration must be given to connectivity within language (relations within and between phonetics, grammar, and semantics). Sign analysis, however, must also consider the ways in which language is converted into acts of speech grounded in real events.

But the argument begs the question: How are individual speech acts hooked up to broader domains such as text and social context? Ricœur's answer is that text and context intervene by *filling in the missing parts of sign and discourse and determining the directions of meaning* (ibid.: 67–9, 79). Inspired by the works of Benveniste, Frege, Richards, and Beardley, Ricœur's argument is that the contextual coordinates of acts of speech have a direct impact on the actual selection of primary meanings attached to signs. Information derived from context contributes to the reduction of polysemy and the avoidance of undue imprecision and systematic equivocation. Only those meanings that fit the total context of sign utterances will be retained, those matching the concrete circumstances of *la parole* (ibid.: 82, 96, 131). To this contextual selection principle Ricœur adds *the rule of plenitude*: Connotations and secondary meanings may be conveyed through figurative language provided they are permissible and compatible with first-degree references and their contextual coordinates.

Ricœur notes that contextual mechanisms do not simply force a limited set of principal and subsidiary meanings to be chosen from pre-established codes consisting of catalogues of agreed-upon associations. After all, the interaction of signs and context can produce original effects and connections, as is the case with unusual amalgamations of words and situations and the creative deviations of metaphor. The malleable associations of sign and context thus account for the genesis of new meanings and novel perspectives on reality and the world as we experience it (ibid.: 124–7). While manipulations of signs-in-context can be found in literature, history in general is also an important source of semantic and hermeneutic innovation. The changing conditions of natural and cultural history require indeed that new objects, values, and social forces be signified through creative transformations and productions of language. Deviations from fixed norms and conventions of the Code are historically inevitable. Language is never so "formal" and "systematic" as to preclude flexible adaptations of signs in context.

Ricœur argues that when acting jointly, the principle of contextual selection and the rule of plenitude permit the act of metaphor to perform its central task, which is to convey all acceptations already codified plus one, "that which will rescue the meaning of the entire statement" (ibid.: 131). John's "Lamb slain" metaphor appearing in Rev. 5.6 thus carries a plenitude of images conventionally tied to young sheep sent to the slaughter (meek, humble, sacrificial, etc.), to which it adds one extra meaning justified by the broader semantic and hermeneutic context, namely, Christ the Saviour as portrayed in the New Testament Apocalypse.

The "*context theorem*" propounded through the Ricœurian advocacy of semantics and hermeneutics undermines all assumptions of a self-autonomous intellect and the aesthetic closure of the Code. But notions of context intervening in text (or language) pose several problems. First, Ricœur tends to equate the broader linguistic context with the grammatical sentence that circumscribes the meaning of each verbal utterance. Allowances should be made for a more comprehensive understanding of semantics, to include the broader script, genre conventions, and cultural parameters that intervene directly in making sense of words and sentences in their proper context. Thus the "Lamb slain" sentence of Rev. 5.6 must be interpreted in the light of other chapters and verses of Revelation, the use of animal symbolism in sacred texts, the appearance of comparable imageries in pagan cults of the first century, and so on.

Secondly, and perhaps more importantly, the concept of contextuality is often equated with all factors exogenous to the sign process and acting essentially "from outside." Things viewed as "contextual" end up forming a hodgepodge of real-life information missing from the text, "extralinguistic background" circumstances to be foregrounded for interpretive purposes. The problem with this context theorem is that someone's variegated background material (facts of individual, social, or natural history) is bound to be someone else's object of study and centre of attention, with literature and text thrown into the background. "External" circumstances, say, the economics and politics of John's apocalyptic vision, are then carved into a subject matter requiring meaningful structure and depth (e.g., Roman imperialism), with text acting merely as surface signs of historical struggles over the exercise of power.

An alternative to this text/context or sign/situation approach lies in the practice of transdisciplinarity, hence hooking up in-depth analyses of two or several domains, for example, text and political economy. This strategy does away with disciplinary assumptions of "residual contextuality." It acknowledges the complexity of each field and the requirements of intra- and inter-regional analyses of culture, society, economy, and polity. Contextuality gives way to a dialogue between fields of study. By admitting more than one focus of analysis for the same phenomena, transdisciplinary studies go a long way in avoiding the principal fault of "contextual reductionism": they do not relegate the non-linguistic to background "filling" consisting of residual *ad hoc* factors intervening in semiotic space.

But solutions often carry problems of their own. Notwithstanding its good ecumenical intentions, transdisciplinarity is no challenge to commonplace divisions established between the "inside" and the "outside." More often than not, it leaves intact all conventional distinctions posited between language and the non-linguistic, or between signs and the "reality" or the "world" we live in. This raises a third problem, which concerns all "immaterial," quasi-metaphysical notions of the sign process. I am referring to conceptions of sign activity that posit differences between the mental or the meaningful and whatever lies outside the realm of knowledge and thought, be they products of the senses, the body, the environment, real behaviour, concrete practices, experiential reality, the phenomenal lifeworld, or the material conditions of social existence. None of these conceptions is self-evident.

By way of illustration, Husserl's phenomenology is one influential formulation of the Great Divide that separates the subjective and the meaningful from the objective and the material. As Derrida (1981: 31-3) notes, phenomenology posits a linguistic mediation between two things. On the one hand, there is this layer of "pre-expressive" experience or perceptive intuition; on the other, there is pure meaning or consciousness signified by language yet lying outside the jurisdiction of the signifier, existing in its own self-presence. The end product of this work of mediation is a "re-presentational" view of the labour of language. Signs serve to translate, express, communicate, or incarnate presemiotic ideas or realities that evade or transcend weavings of the sign tissue. Pre-expressive experiences and thoughts are not subject to the spacing of letter-like or gram-like signifiers, spacings marked by the trace of other letters and grams.

Derridean grammatology takes a different view. It holds that connective traces engendered through *différance* are internal to all moments of signification and constitute meaning itself and the realm of the signified. "The presumed interiority of meaning is already worked upon by its own exteriority" (ibid.: 33). Should we conclude from this that sense generated through signs can be detached from extralinguistic activity and products of the senses? While Ricœur adopts a view of semantic and hermeneutic contextuality that precludes basic properties of sign activity, the Derridean tendency to eradicate all "representable" activity lying outside the letter-like weavings of *différance* is no less problematical. Subsuming all human activity (or knowledge thereof) under text and literature, if only by way of analogy, does little justice to factors of "natural history," or whatever it is that exists outside language, society, and the individual.

In its own way, grammatology maintains the conventional divisions between natural and human sciences. It fails to address the nomadic fluxes and lines that push back all frontiers separating body from sign and brain from mind. In the end, both Ricœurian and Derridean perspectives share a common assumption: though sense (linguistic, inscriptive) may eclipse or interact with the senses (physical), one remains profoundly different from the other.

Making Sense

The position advanced in this book differs from Ricœurian hermeneutics and Derridean grammatology, and questions all presumed divarications of internal meaning and external context. Sign events are not mental events intersecting with sensorial phenomena. While taking different forms, signs are products of "sense" defined as bodily faculty (sense of hearing), meaning (the sense of a word), judgment (being sensible), reason (common sense), emotion (being sensitive), awareness (to regain one's senses), and memory (sense of time, cf. *3-D Mind 3*). My contention is that interconnections between these different facets of sign and sense production are so dense and pervasive that all distinctions between *logos* and *physis* must be abandoned. Through multiple sense activity, various planes hook up. They do so by virtue of everything that is shared and communicated within and between the sign corpus and the body social and organic. This is to say that the distance that lies between sign and situation is synaptic in its own right. It constitutes a space where some neural signals are converted into sign formations – signals configured and processed in such ways as to receive attention and "make sense."

Meaning, judgment, and emotion involve neural signals no less than organic feelings and communications of the autonomic system. But they also represent chemical and electrical activities that apply particular forms of attentionality to pattern and order received and transmitted via subattentional pathways of the body and brain. Vague expressions of "background

circumstances" or "text responsive to context" are no substitutes for studies of sign-sign and sign-signal reticulation, a host
of connections inherently physical in the sense of being processed via the nervous Sr system. Signs and signals belong to a
universe where the weavings of "sense" and "synapse" ignore
all rifts between conception and perception, meaning and experience, psyche and body, word and world.

Divarications between inside-mental signification and out-
side-physical stimulation are a legacy of the metaphysical tradition. To paraphrase Derrida, we could say that the interiority
of meaning is already worked upon by its (physical) exteriority.
Or is it the other way around? Could it be that the outsideness
of meaning (meta-physical) is already shaped by chemical and
neurological activities happening inside the brain? Should we
not question the internal/external terminologies, including
those that serve to distinguish the sign logic of LH activity from
the tactile and sensorial features of RH processing? Jakobson
(1985: 167) once suggested that the RH specializes in perceptions of extralinguistic activity, including non-speech sounds
and noises in general. By contrast, the LH processes inputs and
outputs of meaningful linguistic activity. We have seen that
things are more complex. The RH can *construct meaning* out
of internal or external stimuli without linguistic input or output. The notion that sense or meaning should be confined to
words is therefore faulty. Conversely, linguistic operations of
the LH are dependent on internal and external stimulation combined with motor activity and synaptic operations of all sorts.
The notion that meaning and sensation are distinctive functions
localized in different areas of the brain is without neuropsychological foundation.

Signals are not mere sensorial responses to external stimuli,
to be given meaning through the intervention of "metaphysical"
intentions or thoughts and words to express them. As Merleau-
Ponty (1962: 3, cf. 7) remarks, signals are not pure physiological sensations produced by "the experience of an undifferenti-
ated, instantaneous, dotlike impact." They are not instant
snapshot events transmitted sequentially, through chain-like

reactions of nervous functioning that simply copy and communicate information of the senses. Rather, signals are shaped by
central nervous operations that organize reality, introducing
some "rule of conformation" into views of objects and reality
otherwise chaotic. Through this rule, a lamp can be apprehended and recognized as a full object without the subject actually perceiving all sides and facets of the object. The lamp
"makes sense" by virtue of a "practical synthesis" that incor-
porates possible or necessary aspects not made present to the
senses. Properties of "virtual reality" are fed into an overall
unity that is "sensible" rather than ideational or conceptual.
The "real object" may exist only if perceived, yet it exists well
beyond what is actually perceived. "It" is thus a phenomenal
reality "given as the infinite sum of an indefinite series of per-
spectival views in each of which the object is given but in none
of which is it given exhaustively" (Merleau-Ponty 1974: 199).

The rule of conformation also means that signals are processed within broader connective fields that place each and
every perception in the middle of multiple conformations (ibid.:
4, 8, 14); rather than being a perfectly solitary object, the lamp
may be in a countryside house spotted at a distance by a man
lost in the forest. Merleau-Ponty adds that these multiple conformations are not merely of the informational sort. All perceptions and productions of meaning also have a motor
accompaniment, "incipient movements which are associated
with the sensation of the quality and create a halo round it."
This means "that the 'perceptual side' and the 'motor side' of
behaviour are in communication with each other" (ibid.: 209–
10). A musical note is thus "merely the final patterning of a
certain tension felt throughout the body" (ibid.: 211).

The same can be said of colours. They are never simply apprehended as discreet qualities that occupy delimited positions
within fixed codes, qualities that combine with other codes to
produce the configurative effects of perception. Rather, a colour
such as red is always part and parcel of an overall "field or
atmosphere presented to the power of my eyes and of my whole
body" (ibid.: 210). There is no redness without a sign event, a

"bodily attitude" or intention that hinges on all those field circumstances that determine the existential colouring of red. The sense of redness may emerge from a scene of violence involving the spilling of blood, but it may also be tied to an erotic invitation signified through facial makeup or red lights. Alternatively, feelings of danger and anxiety may be signalled through a red alert, a scenario that differs from sentiments of shyness expressed through blushing. Through colours of determinate events, the whole body surrenders to a "particular manner of vibrating and filling space known as blue or red" (ibid.: 212, 215).

The rule of conformation involves a synthesis of physis and semiosis applicable to all productions of the nervous sign system. But it is not the only rule intervening in the order of sense. As I argued in *3-D Mind 1*, effects of convergence are obtained with support from the opposite principle, the *rule of fragmentation*. Compositions of signs and signals can be put together only because they can be dismantled. All assemblages feed on fragments released and retrieved from other assemblages. The rule of conformation implies that a recognizable sound-image such as the interjection "ouch" (triggered by the burning of one's hand) will be accompanied by a specific breathing pattern and particular movements of the larynx, the mouth, and facial muscles. All signs and signals converge on the expression of pain. The rule of fragmentation, however, makes it possible for physical accompaniments (pain sensation and respiratory rhythm) to survive the fading of the sound-image "ouch." They are freed from one "practical synthesis" and serve to inform or trigger a new assemblage (moaning, shaking the burnt hand to reduce pain), thereby allowing the previous composition to both continue and evolve.

One critical source of fragmentation comes from the way in which attentionality is unevenly divided between component parts of a given assemblage. For instance, the word "ouch" may be uttered in a loud voice, while breathing in gasps is performed automatically, without noticing. Verbal signs and respiratory signals are part of the same event, yet they are fragmented along attentional lines. One of the most important

mechanisms whereby signs and signals are assembled, disassembled, and reassembled lies in these differential allocations
and shifts in attentionality. Attentional economics account for
movement in all "syntheses of sense."

But we must remember that attentionality is the tip of an
iceberg called the body, the larger intelligence of the full nervous
system. While they are a visible manifestation of brain activity,
attentions of the "mind" are constantly supported by connective
activities that receive little notice, if any at all. Signs and signals
are subject to variations of foreground and background attentions, measurements of awareness that play a critical role in all
acts of semiosis. Levels of attentional determination are integral
features of reticular sign-signal activity. They give shape to both
lines of conformation and fragmentation woven into products
and motions of body and language.

The sign-signal economy is driven by quanta of awareness
that preclude simple either/or allocations of attention. The processing of sense should be understood as a non-digital faculty
that leaves room for ambiguous or fuzzy syntheses of signs and
signals of the manifold. The brain is not programmed to make
simple-minded choices between consciousness and the unconscious, leaving no room for variable intensities of attentional
light and shade. In his critique of theories of well-bounded
objects of perception, Merleau-Ponty ignores these chiaroscuro
effects of attentionality. His view is that the notion of inattention
is of little use in explaining the phenomenon of perceptual ambiguity. He concludes from this that the notion of attention is "no
more than an auxiliary hypothesis, evolved to save the prejudice
in favour of an objective world. We must recognize the indeter-
minate as a positive phenomenon" (ibid.: 6). Fuzzy perception
is part of how the body is situated in the world, as opposed to
a failure of knowledge (cf. Hammond et al. 1991: 167). But
what is indetermination in the first place, we may ask? Is it not
the lower end of mechanisms of configurative noticing, a fuzzy
attention that may sustain the "intentional part" of some perceived meaning or meaningful perception (Merleau-Ponty 1962:
13)? Merleau-Ponty argues correctly that indetermination does

not simply result from some "attention deficit" blurring our perception of determinate objects. But the malleability of attention should not be reduced to a pathological condition. It constitutes rather a powerful instrument of perception.

Three basic conclusions follow from our Sr (sign-signal reticulation) theorem. One is that perception should be granted primacy over the intellect, as in the writings of Merleau-Ponty. This is not to say that knowledge should be reduced to sensation and that sensation is the only thing we see, as Pascal once proposed. Rather, perception should be understood for what it is, a "nascent logos" that never escapes the world we live in, a knowing process that is no less sensible than the sensible and no less intellectual than the intellect (Merleau-Ponty 1974: 203, 210).

Secondly, Merleau-Ponty is also correct in arguing that a sign is essentially diacritic and always appears as a trace. Signs are profiled against other signs. Language thus consists of differences without selfsame terms, and meaning lies in the interval between words (Merleau-Ponty 1964: 39, 42, 88). By implication, meanings generated by the intellect are indistinguishable from the operations of *la langue* (conceived broadly, to include semiotic practices other than speech and writing). Since "meaning is the total movement of speech, our thought crawls along in language" (ibid.: 43). The same reasoning, however, should be extended to sense understood as *meaning, judgment, feeling, and sensation*. Sense itself is a relational fabric, and meaning is shaped by the intervals between signs and signals, linguistic or not. This lateral liaison principle means that speech acts are always situated against broader processes of synaptic communications and signification, transmissions constantly turning and folding back upon themselves. Speech utterances end up being merely folds and interfolds in greater fabrics of "sense."

Thirdly, a basic rule of attentional economics is that not all sign-signal connections need to be voiced or spelled out through the attentions of *la parole*. What Merleau-Ponty says of language applies to all practical syntheses of "meaning and sense." All such syntheses are inherently opaque and full of secrets, ruled by attentionalities interwoven with threads of silences, allusive indirections and traces of the oblique, the lived-in, and the unnoticed. Writers may think that what they write is an elaboration of signs brought out into the open. In reality, they are like painters who generate meaning through the silent world of lines and the vagueness of colours. Writing echoes the subtle hues and lines pervading all "mutable" inscriptions of meaning. Signification always "contains" (conveys/withholds) a mute art (ibid.: 44–6).

Figuratively Speaking

Our sign-signal connectivity theorem would be all the more useful if it could account for variations in the art of symbolling, variations other than purely circumstantial. The reticulation theorem is of no great consequence if it simply means connecting sign to circumstance. Consider metaphorical namings of stars and constellations, say, from a substitution-theory perspective. When looking at asterisms on a map or stars in the sky, one may use terms such as Scorpion or Taurus with full knowledge of their "deviational" properties. No "real" scorpions or bulls are intended. The terms are used in the absence of any alternative word or descriptive paraphrase that could restore the meaning intended; signs that metaphorize in this manner cannot be understood with reference to other signs, at least not the kind that can be communicated with brevity. The context is sidereal and yet the language is zoological. In keeping with this analysis (based on substitution-theory), a proper name marker, the capital letter B in Bull, indicates a deviation from common animal lexicology; thus "look at the Bull" does not equal "look at the bull." What we have here is a case of denotation achieved through metaphorical loan, one might conclude.

From a Ricœurian perspective (Ricœur 1977: 74), we might say that contextual references and semantic intentions are needed to "make sense" of these metaphorical namings of astrology and astronomy. I know what "scorpion" means if the person uttering the word happens to be pointing at stars on a celestial map. Meaning is determined by situation and not merely by substitutive convention highlighted through upper-case lettering. The principle is so simple as to be self-evident. The problem with this approach, however, is that lines on a map or configurations in heaven are no less symbolic than the word "scorpion." Meanings on a map or in the observable sky are not referential material directly perceived by the senses and "re-presented" in figurative language.

The *sign-situation theorem* is actually so weak as to neglect some basic facts. When words hook up with perceptions of dots and configurations on a map or in the sky, they "synapse" not with outside reality but rather with signs and signals processed into syntheses of "sense" – events comprising both sensation and signification. Verbal signifiers are crisscrossed with acts of meaningful perception; signs and signals are interwoven into a living tissue simultaneously physical and communicational. Thus an eye focusing on something is no less "meaningful" than the verbal utterance associated with this event. Nor is it more physical, situational, or referential. The "scorpion" utterance points not to the mind adding intelligibility to sensation, a verbal act more "mental" than the visual recognition of a con-stellation appearing in heaven. The word "scorpion" is *rather just another cog in the wheel*. If signs and signals can connect to a referential reality, they will do so essentially through com-plex connections established among themselves. When ade-quately understood, the sign-signal reticulation process spares us the necessity of grounding "sense" in pure referential exter-nality, or reducing meaning to pure internalities of the mind.

A better case of referentiality unmediated by the sign-signal process could perhaps be made for descriptive utterances such as the personal "I" pronoun. The sign of a person speaking, this nominative term seems to serve a purely auto-designative function. It appears to refer to the speaker and nothing else. The same could perhaps be said of the *present tense*, which denotes being in the absolute present and the immediacy of the moment of speaking, hence words of the here and now. This is the view held by Ricœur. But the argument is terribly weak. For one thing, the "I" sign belongs to a grammatical code that

positions signs of the self in relation to other personal pronouns, doing it in ways that will change from one language to
another. Pronominal codes are not everywhere the same. The
selfsameness and self-referentiality of the subject calling itself
"I" is subject to variable systems of individuating differences,
distances (as between neuter and gender, male and female, singular and plural, reference and address, etc.) that precede and
make possible all constructions of the "I" and the self. Like all
other signs, notions of the self are affected by sign differences
that intervene like ants entering and leaving through fractures
of the "I" (Deleuze 1994: 277).

Effects of "signaptic" connectivity apply to "being in the
present" as well. In Spanish, the English expression *I am* may
be conveyed in one of two ways. It can be expressed through
the word *soy* uttered without the pronoun (why this self-efface-
ment vis-à-vis the verb?). Alternatively, the verb *estar* can be
used to denote being with implications of short duration and
transiency. While taking different forms, the present tense will
also make sense in contradistinction to other tenses built into
a particular grammar. Thus there are no universal rules that
determine how the present differs from the past, the future, the
subjective, the conditional, and so on. In the end, nothing
escapes the gaps and weaving of signs and signals, not even (or
especially not) the grammatical formulations of "being in the
present" or self-presence in the "I."

But what about *homonymy, synonymy, polysemy,* and *lexi-
cality*? Are they not good examples of gaps and connections
between sign and external reference? Ricœur answers again in
the affirmative. In the case of homonymy, one name "refers" to
one of various "things." Verbal signs that have different senses
and references may have the same pronunciation and even the
same spelling: a "pupil" is either a child receiving tuition or an
aperture in the iris of the eye. The intended meaning depends
not on the word used but rather on the referential interaction
of sign and context (Ricœur 1977: 112–13). With synonymy,
referential intentionality takes on a different twist. Meaning
remains constant while signifiers vary; someone showing up at

forty-five minutes after ten is as punctual as someone arriving at a quarter to eleven. This goes to show that objective reference can survive sign substitutions. There is also polysemy. One name activates several meanings and references, those that can survive the siftings of textual and contextual relevance. Again cues provided by extralinguistic circumstances and other indices of language (e.g., the word "pupil" uttered in a classroom context, without anatomical implications) are needed to arbitrate the multiple acceptations of a particular utterance (ibid.). To these variable connections between sign and reference can be added shifts in lexicology. The shifts in question involve signs losing or gaining references and references losing or gaining signs, thereby altering conventions of the ruling lexicon (ibid.: 115–17, 123, 127).

Ricœur's sign-situation dialectics appeals to common-sense notions of word and meaning. All the same, it oversimplifies the nature of contextual variables, treating them as external "circumstances" that share little with signs used to signify them. The attentions of sign and signal reticles, connections that boil down to neither mental nor physical referentiality, offer a more plausible account of manipulations of word and meaning. Consider again the teaching of metaphor in regards to how signs interconnect and receive unequal attention. Readers are reminded of our previous argument: metaphors involve sign substitutions obtained through intra- and intercode borrowings and loans. Given this reasoning, "category mistakes" or "lexical faults" can produce metaphorical effects on three conditions, all of which pertain to the process of uneven attentionality.

First, while the metaphorized sign (e.g., Christ symbolized by the Lamb) may not be spelled out, at least not in the immediate text, the sign-metaphorizer (the Lamb) must be foregrounded. The connection between the spoken and the unspoken must be noticed in the same breath. Otherwise the metaphorical effect cannot be obtained. A metaphor performs its task provided that greater attention is paid to (a) one sign in relation to another (foregrounding the substituting lamb, backgrounding the substituted Christ) and (b) connections between the two.

Second, predicative parallels projected via the language of metaphor must be acknowledged, if only implicitly. In the Lamb slain imagery, attributions of the lamb (e.g., meekness) are passed on to the Saviour through what Ricœur calls the *plenitude principle*. The metaphor is an invitation to explore all relevant similarities; to metaphorize is to perceive resemblance, says Aristotle (ibid.: 24). It should be emphasized, however, that not all possible attributions of the metaphorizer are preserved in the act of metaphor (ibid.: 107). In keeping with our attentional theorem, some linkages are triggered whereas others have to be bracketed (the Lamb slain imagery is not meant to trigger sheep odour implications, for example).

Third, the "lexical fault" conveyed through metaphor must be constructed with full cognizance of the deviation or misuse of the metaphorizing sign or code. Christ is not really a lamb; nor can a "queue" of people standing in line be wagged. This means that for a metaphor to work, both similarities and differences must be simultaneously attended. Without attention to dissimilitude, the metaphor turns into sheer nonsense or mere tautology and loses its meaningful surprise effect (ibid.: 82). A metaphor avoids lapsing into an abuse of identity by drawing subsidiary attention to the "signaptic gap" that keeps similars at a distance.

The three conditions outlined above converge on a basic principle that crosscuts the sagittal interlacing of RH similarities and LH differences: namely, the axial projections and attentions of signs (and signals) rank ordered and unevenly distributed in semiosis. We have seen how these attentional mechanisms intervene in the language of metaphor. But they are also at work in the production of homonyms, synonyms, polysemy, and shifts in lexical history, effects that play on linkages between "signs in context." To use a terminology that avoids the vagueness of "context," we might say that all devices of rhetoric involve reticular attentions operating at two levels. At one level, we have malleable connections between soma signs – signs effectively attended or noticed. The other level consists of communications proceeding through autonomic signals, Sr activity unattended via the noticing of language and syntheses of the senses.

Homonymy represents a particular form of sign-sign/sign-signal connectivity. It involves one phonetic reticle that permits at least two possible sets of connections and a requirement to choose according to the broader sign-signal assemblage. Thus the word "pupil" can be combined with signs pertaining to either the anatomical field (eye) or the social domain (tuition). Speakers and listeners will depolarize one sign reticle only, using indices provided by other verbal utterances and/or practical syntheses of the senses. With homonyms, zero-degree interpretive flexibility is therefore imposed upon connections between sound and image. One choice precludes the other. The absence of flexibility that comes with a homonym means that no attention is required when making the choice. While homonyms imply choices between sign connections, the either/or choice to be made depends on a host of signals processed by the brain and the senses, *with or without cortical attentionality*. The "mutual selection of acceptations of semantically compatible meaning is most often effected so inconspicuously that, in a given context, the other inappropriate acceptations do not even cross one's mind. As Bréal remarked already about this, 'one does not even go to the trouble of suppressing the other meanings of the word: these meanings do not exist for us, they do not cross the threshold of our consciousness'" (ibid.: 131).

A sign-signal assemblage involving a doctor who remarks on the size of a patient's pupil during a medical examination points to a medical-anatomical machinery. This assemblage precludes a social-educational composition centring on the "pupil under examination" imagery. The preclusion is arrived at mechanically, through automatic inattention to signs other than those of the medical field.

A homonym triggers a reticular obligation: paying heed to appropriate associations and ignoring all inappropriate ones. By contrast, a *synonym* presupposes some degree of freedom and arbitrariness in the act of reticulation – not having to justify or take notice of the fact that a particular sign is attended instead of another. The subtractive imagery of a quarter to eleven can be substituted for the additive expression of forty-five after ten, without the substitution requiring any explanation. The freedom is not absolute, however. While they do not impose binary choices, synonyms usually betray preferred associations and subtle choices, without full awareness of the choices involved. For instance, when the word "expiring" is used in connection with evocations of death, the act of breathing out is singled out as a key symptom in the cessation of life. For some reason, signs of breathing are then preferred over other equivalent expressions such as "life fading," "the last journey," or "the final rest." By selecting one particular word or expression instead of other possible synonyms, a particular slant is imposed on the composition at hand. What Jakobson (1985: 119) says of half-synonyms may thus apply to all synonyms. Depending on how they hook up with other signs and signals, synonyms may offer choice between slightly different attentions.

The slant suggested by a particular synonym can point to subtleties of mood and associated signals. As we all know, the negative affect tied to a half-empty glass differs from the positive mood of a cup portrayed as half-full. The same can be said of time gone by at ten forty-five as opposed to the quarter of an hour that remains before eleven. Although apparently referring to exactly the same "thing," two equivalent signs may accommodate sub-attentional discriminations of affect and tone. When brought together in the same script, synonymous configurations can also produce particular attentional effects. Think of the exaggerative piling-on effect of "a ceiling painting portraying the uplifting scene of the son of the Most High ascending to the lofty heavens." In this scene, signs playing the role of synonyms serve to add on connections of the "lofty" sort and sustain the attention vis-à-vis the corresponding field of signs and signals, with effects of iterative amplification. The intent is to avoid a purely instrumental language that offers no sense of amplitude and will capture no one's attention.

Polysemic reticles, words that connect to a plurality of signs all at once, show an even greater ability to enhance the freedom of Sr attentionality. Compared to metaphor, polysemy entails a looser management of the sign-noticing effect. The focus is on a particular imagery entailing a number of connections that can be left "semi-attended," on a "lower synaptic frequency," as it were. Also polysemic linkages are usually many, without there being one preferred set of metaphorized signs or signals. To use a biblical example, the polysemic ramifications of God's Love Feast form multiple lines radiating from a central motif, as opposed to taking the form of a two-sided connection as in the Lamb-Christ metaphor (a motif with polysemic implications of its own).

Contrary to most theories of metaphor, polysemy is not for all that symptomatic of the vagueness or pathology of meaning. Nor is it symptomatic of the infinite freedom of signs escaping or deviating from the strictures of language. In its own way, polysemy is a contribution to the orderly ways of language. As Ricœur (1977: 127) remarks, multiple "meanings" conveyed through polysemic signs reinforce the economy and flexibility of the sign process. One word can be used to activate plural connections if and when plurality is in need of both attention and verbal parsimony. Moreover, polysemy can be used to organize signs into fields of interconnected acceptations and predications, fields that share common boundaries and require organization. For instance, polysemic evocations of the Last Supper emerge from a choice made in favour of a privileged sign – the Last Supper instead of the Eucharist (Greek, giving of thanks) or the Agape (Love Feast). The larger field thus receives a particular slant. The Last Supper expresses a preference for images of sacrificial and festive feeding over those of thanking or wedding. Paradoxically, polysemy places restrictions on the fuzziness of signs. It constitutes and connects the explicit and the implicit through particular attentional arrangements. In the words of Heidegger (1968: 71), the "multiplicity of possible interpretations does not discredit the strictness of the thought content ... Rather, multiplicity of meanings is the element in which all thought must move in order to be strict thought."

Polysemy is orderly in another way: it shows responsiveness to the attentions of nonlinguistic signals. A man can exploit the pejorative connotations of the metaphorical "ass" insult on condition that his object of attention is not a four-legged donkey standing in his immediate field of vision. Signals of a sensory deviation from strict animal designations are required for animals and personality attributions to be hooked up with pejorative intent. In other words, polysemy involves yet another assemblage of signs and signals, as opposed to a mental act mediated by words and circumscribed by external context.

Finally, changing lexicologies introduce the time factor into Sr activity. With the passing of time, new words can be applied to innovative sign connections and attentions: for instance, the word *e-mail* to acknowledge and signify a novel assemblage of electronics and postal activity. Alternatively, lexical change may consist in extending an older sign to a previously unconnected field. The word *reticle* as used above, a term drawing new attention to parallels between biology and semiotics, is another product of shifts in lexical montage.

Variations in how attentionality is granted to sign and signal connections go a long way in accounting for the effects of rhetoric. Attention is not an either-or option but rather a malleable function that determines the precise shape and contour of language. Graded movements between foreground and background (and also between ground and underground) are particularly important in this regard. They are essentially responsible for the ongoing management of all signs and signals that can be brought to bear on any matter, script, or task at hand. They also enable us to understand how semiotics can be reconciled with the semantics, the hermeneutics, and the pragmatics of nervous sign activity.

Concessions to Literality
and Grammar

What are we to make of signs interpreted literally, technically, descriptively, or denotatively? Do they link up with other signs in the same way that metaphors do? Is it in their power to connect directly to objects and phenomena belonging to reality, without multiple sign connections and the uneven measurements of attentionality? Linguistic realism and common sense tend to be on the side of the "simply denotative" attitude. With proper denominations, the question is therefore whether the sign is "simply" accepted or not. As Benveniste would say, does it signify or not? The counterargument to this is that all words are in some way figurative and therefore metaphorical. Even the word *figure* as applied to speech is figurative. The word was said originally of bodies, or humans and animals considered as physical and apprehended with respect to the limits of their extension. A "figure of speech" thus evokes visibility and spatiality, a quasi-physical exteriority involving contour, feature, or form (Ricœur 1977: 143–5). Similar comments apply to metaphors. We cannot speak of them without doing it metaphorically. As Ricœur (ibid.: 17ff., 137) points out, the Greek combination of *meta* (over) with *pherein* (to bear) suggests a change of location and the act of transportation. When speaking of metaphor, we use a spatial movement metaphor.

The two positions outlined above are diametrically opposed. The choice is between accepting or rejecting the great divide between the figurative and the literal. The issue has been the object of considerable epistemological debate. When attentional processes are factored in, however, the two positions are no longer incompatible. On the side of figuration, we must recognize that all signs make sense within webs of signification; no word or sign can be isolated from the broader networks of language. The complexity of these networks is such that every sign has built-in classificatory, etymological, and lexical specifications not directly dictated by objects or events. On the side of literality, however, we must recognize that the "powers of speech" include a vital capacity to focus, zeroing in on narrow sets of connections. When this occurs, all other possible weavings are bracketed, unattended, and forgotten, whether by means of polarization or hyperpolarization. When figures and metaphors are discussed, no attention needs to be drawn to shape or movement, irrespective of the spatial metaphors originally built into these words. Likewise, when we speak of logical forms, no contours or container are intended or attended (ibid.: 89).

The principal lesson of literality is that all productions of "sense" require some quanta of inattention, measures permitting us to ignore all things that need not be attended. This implies that literal meanings can be obtained through concentrated attentionality and maximum inattention to everything else, hence focusing on the narrowest set of relevant connections. The sharing of attentionality through metaphor or other means is simply ruled out. When asked to turn left at First Avenue, few will ask themselves if they are being pressured into converting to socialism somewhere down the road. Similarly, when asked to look at the Milky Way, few will engage in a synaptic facilitation of linkages between dairy products and the galaxy. While not a negation of the weavings and framings of language, "literal focusing" is a vital attentional option to be used and deployed whenever required, *with or without emotive investment or the awareness thereof.* (Contrary to what many suggest, emotivity is not a prerogative of the language of evocation; see ibid.: 146.) Theories that purport to undo all distinctions between literal and figurative names, or the denotative and the connotative, as in the work of Gadamer (1994: 432; cf. Ricœur 1977: 23), play a critical role in challenging naive notions of sign-object representation. They nonetheless neglect a fundamental principle of semiosis: the uneven allocations of attention.

While differently focused, effects of figuration and literality are not boxes into which signs can be neatly categorized. Semiotic attentions are not quanta of meaning fixed into immutable conventions and codes. Rather, they constitute dynamic measurements that constantly shift and move. Attentional movements can transform figures of speech into highly focused denotations, metaphors that die away but also generate new lexical systems. Shifts of this sort are essential to lexical innovations that deviate from old conventions and create new frames. The process by which logic is disrupted is thus the same as that from which logic proceeds. The metaphoric "that transgresses the categorical order also begets it" (Ricœur 1977: 24). Violations of established codes engender new codes and connections within and between fields. This is to say that acts of sign deviation and transgression are never purely negative or chaotic. When used in non-conventional ways, metaphors are subject to some reduction of deviation whereby new norms are used to mark off the boundaries and regulations of innovative terminologies (ibid.: 149).

Metaphors often feed into this lexical innovation process. However, they can also be used merely to foreground possible linkages previously unexploited, making them explicit without altering existing codes. The phrase "old age is a withered stalk" may be poetically new to some, yet the imagery of lost bloom is not an attack against better-known imageries of old age. Just as literal attentions can break new grounds with the help of previous metaphors, so too metaphors deviating from literal meanings can revisit themes built into old codes.

An example of a realignment of metaphors and codes can be found in New Testament imageries of scorpion and lamb, fall and spring, demons and gods. Briefly, in Revelation animal symbols break away from dominant astromythical thought. The Book of Revelation shifts the reader's attention away from conventional connections between animals on earth and idols in heaven, pagan beliefs deemed objectionable enough to be downplayed and forgotten. The prophet brings down the older pantheon by turning animal-shaped stargods into zoological metaphors. Divine spheres in heaven are treated as mere sign manifestations and figures of speech. In lieu of being confused with the son of vernal Aries, Christ appears in "the likeness of the appearances" of a Lamb slain. By undoing hegemonic codes of "pagan" inspiration, John's metaphorical innovations introduce a new frame of reference, with rules and constraints of its own. His language of prophecy (substituted for signs of divination) heralds a regime focused on the Great Divide between body and spirit; it announces the reduction of animal and heavenly bodies to sign-manifestations of Logos understood as the Verb or the word of God.

Animal metaphors in Revelation do not merely break older conventions through associative innovation. They also create new confines for language, boxing the sign process into *either* visible sign-manifestations *or* their invisible Sign-Makers. Through a lavish deployment of apocalyptic metaphor (and related developments in theology), signs of astrology and divination give way to signs of prophecy (see Chevalier 1997).

The attentions of metaphor can be innovative and orderly at the same time. Although highly supple and malleable, meanings conveyed through metaphor will not slide chaotically or shift ad infinitum, without the siftings and screenings of attentionality. To paraphrase Ricœur (1977: 130), while a sign connection may be a *plural* identity, it is nonetheless a plural *identity*. Even when taking on the functions of polysemy, multiple sign pathways are not open-ended tissues deviating from all rules of logical and attentional reduction. Nor are poetry and metaphor the antinomy of fixed lexical catalogues and storehouses of commonplaces. The plurivocality and mutability of meaning and its sensitivity to context in living languages should not be opposed to the rigidities of artificial languages, as in Jakobson. Even when "eccentric," all acts of signification are "framed" by configurations of signs and signals that bring orderly attentions into motion. Every sign weaving is "a limited, rule-governed, and hierarchical heterogeneity" (ibid.: 130).

As with literality, grammar is another constraining feature of language that can be pitted against the freeplay of metaphor. Ricœur (ibid.: 131–2), however, reminds us that signs and metaphors entail predicative effects resembling those of full-grammatical propositions and comparative statements. Metaphors are abbreviated sentences, as it were. But there is a problem with this counterargument. It presupposes that metaphor and comparison are essentially the same. The only thing that differs is the explicitness or implicitness of predication. Yet in reality the latter choice – using or not using grammar to spell out a metaphorical parallel – is exercised for a reason. Sentences can be fully uttered only by virtue of the attention we pay to explicit grammatical vectors and structures. The question is why and when should the speaker bother to convert metaphors into predicative semantics – sentences that provide the broader context of namings and figures of speech?

To say that all metaphors are abridged comparisons is to avoid the question. Extending the attentions of grammar to all metaphors is to use the concept of syntax metaphorically. It denies the distance that lies between predicative associations constructed with syntax and those constructed without it – without the overt expressions and formal requirements of discourse. In the final analysis, the question posed by predication is not how signs negotiate their meaning through tacit or explicit interaction with sentence, but rather, what does grammar do to sign attributions *if and when* applied? Does full grammar add anything to metaphor? When and why should the interlacing of signs and signals dispense with the attentions of grammar (as in painting)?

When signifying resemblance, grammar will generate simple verbal, prepositional, and adverbial connectors such as *resemble, like,* and *as.* Do these vectors matter in the constructions of metaphor, and what do they tell us about the management of attention? Consider the distinction between metaphor and comparison, also known as *simile* or proportional analogy. Two theoretical positions can be taken in regards to this distinction. If viewed from a semantic perspective, with an emphasis on the predicative functions of language, then metaphors may be defined as abridged comparisons. They act as predicative similes expressed without grammatical markers of likeness. According to Ricœur and also Black, a metaphor is a verbal short-circuitry that compresses the full predicative statement, producing an analogy that takes on "an air of identification" (Ricœur 1977: 85, cf. 47, 118). Conversely, if viewed semiotically or from an Aristotelian perspective, with an emphasis on the codifications of sign units and relations, then comparisons may be defined as full-length metaphors. They are substitutions and correlations decked with grammatical ornaments. In this perspective, grammatical ornaments such as *like* or *as* can always be left out, if only for reasons of parsimony, swiftness of communication, or clarity and vividness. The two theories share a common assumption: in the final analysis, the presence or absence of grammatical indices of likeness matters little. If so, why does language oscillate between putting in and leaving out these connective markers? Why should connections by similarity display or eschew an "air" of full identification conveyed through metaphor?

To use our apocalyptic example, does it matter that John spoke repeatedly of "the likeness of the appearances" of a lamb slain, as opposed to simply hailing or naming the Lamb slain and maintaining the abridged zoological metaphor? The answer is that grammar does indeed matter. Better said, it "makes a difference." In what way, though?

The answer to this riddle lies in connections between grammatical markers and signs of similitude and dissimilitude. Paradoxically, the more explicit comparative particles are in the use of four-term proportional metaphors (a is to b as c is to d), the more attention is granted to measures of difference. That is, the more discursive the markers of likeness are, the more attention is paid to the gap that separates the sign acting as metaphor and the sign it metaphorizes. In the case of Revelation, discursive parallels that make use of full grammar entail greater reserve in attributions of closeness, hence warnings against possible abuses of identity and related heresies of zoolatry and astromythology. To simply call Christ a lamb, without the grammatical reservations of likeness and similitude, is to invite predications of near-identity, something the prophet is not inclined to do. Though more economical, the expression can lead to an abuse of identity, a failure to sustain reservations of signs of likeness in a context where temptations of earthly and heavenly zoolatry abound. By contrast, wordy predications of Christ manifesting himself in "the likeness of the appearances" of a lamb serve to enhance operations of partial and selective agreement, the kind obtained through cautious comparison. Implications of distance may be reinforced through the use of capital letters, turning the common noun "lamb" into a proper name that signifies deviation from ordinary zoology. Proper naming through upper-case lettering is another grammatical indication of the logic of full-fledged comparison prevailing over metaphor.

Compared to metaphors, expressions of "comparative reserve" put greater stress on the injunction *to connect but not to confound.* Richard is correct in saying that to compare is to connect, but the opposite is also true: comparisons require disconnective activity. While the mind is a connecting organ, it also disconnects and reconnects. It can associate any two things (e.g., lamb and Christ) in numerous ways, which is not to say that all of them are equally acceptable. Some must be rejected (e.g., Christ *is* a lamb) through proper qualifications, with grammatical insistence if need be (e.g., Christ manifests himself in "the likeness of the appearances" of a lamb). Rather than being merely a logical tool, grammar may be used as an attentional instrument that serves to adjust the syncretic amplitude of metaphors in language.

Unlike the Lamb slain imagery, some compositions require the elisions of grammatical attentions. Consider again Descartes's analysis of "*Dieu invisible a créé le monde visible.*" The Cartesian and Chomskian interpretation of this sentence is that deep/implicit grammatical connectors can be activated without being spelled out. Against this principle of structural economy, one should note that the naming of "God the invisible" produces a nearly tautological or definitional effect that dispenses with predicative judgments expressed through full grammatical linkages. The effect may be attributed to reasons other than strictly grammatical ones. To say that God *is* invisible would be to invest speech with the attentions of judgment and the ascertainment of truth, hence the possibility of reversed grammatical constructions lapsing into false beliefs (e.g., God is visible, as in "paganism"). The question is whether or not the sentence should be constructed in such ways as to attract attention to these possible inversions and invite debates regarding heresy. The compressed "*Dieu invisible*" formula answers in the negative. The abridged formula offers the advantage of *appearing* to border on tautology and being less problematical or polemical than a full-length deliberative proposition (God who *is* invisible created a world that *is* visible).

Choices made between denotation, metaphor, and comparison have one basic thing in common: all are choices made between levels of resemblance. What is at stake in all cases is an evaluative measurement of dissimilarity – assessing the relative distance that must be maintained between birds otherwise of the same feather (literality is the greatest distance that can be imposed between a sign and what it could evoke if it were permitted to do so). Attentional tactics expressed through such choices are largely ignored in poststructuralist discussions of dissemination and plurivocality in language. Chomskian and Lévi-Straussian analyses of formal linguistic structures already set the fashion in this regard. Paradigms of structuralist inspiration address issues of deep and surface structures, yet their tendency is to reduce the disparities between the two levels of codification to matters of parsimony and sheer laziness. The unconscious is so lazy that it will do everything to produce effects of meaning, the kind that may be attained without spelling out the underlying rules of phonetic, syntactic and semantic communication.

Structuralism's lazy-unconscious thesis neglects a fundamental operation of semiosis: the *motivation* that goes into specifying the amount of real attentional energy – none, a little, a lot – that various component parts of sign activity should receive. Incidentally, rules established by convention or habit are no exception. They too have to choose between hiding or bringing attention to themselves. They must opt between classical and modern forms of poetry, so to speak. Barthes has remarked that a distinctive feature of classical poetry lies in the cultivation of conventions of formal speech, as opposed to modern forms that allow greater looseness and sheer contingency in verbal artistry, or appearances to that effect (1982: 54–5). The same distinction could be applied to variations in sign activity. All sign reticles may be rule governed, yet some make a point of showing it while others insist on denying it – thereby ruling out rules with tools and techniques to that effect.

Language can choose to signal its own existence. It can do so through metaphorical usages, comparative grammar, rhetorical devices (e.g., euphemism), or the conventions of poetry (e.g., rhyme and rhythm). The attentions *of* language converted into attentions *to* language add something to whatever is being said, namely, "Watch, this is metaphor, poetry or rhetoric!" (see Ricœur 1977: 147). The question again is what motivates rules of speech to "show off" in some situations and to adopt a low profile in others? The question applies to all forms of semiosis, including words, gestures, music, and even architecture. Variations in architectural design can choose to either pass unnoticed or dress up to the nines. The issue is when to deploy a particular attentional strategy and when to forego another. When situated in their own socio-environmental surroundings, Basque houses will speak silently to the taken-for-granted conventions of Basque architecture. If relocated in Paris, however, the same house or simile thereof betrays the imperative hailing (and schematic impoverishment) of "ethnic Basquity" written all over the attraction of a "natty white chalet covered with red tiles" (Barthes 1982: 111). From habits of cultural construction, dwellings can be converted into strategies to capture, monopolize, and manipulate attention. When reaching hegemonic proportions, attentional tactics iterated on a large scale account for much of what Barthes and others view as the interpellations, injunctions, and prejudices of signs frozen into ideology.

In the end, there is more to metaphor and analogy than the perception of similarity in dissimilars, as Aristotle once claimed (Ricœur 1977: 6). Figures of speech draw attention to indices of likeness. But it is also in their nature to generate *variations of syncretic amplitude*, hence shifts in the attentions granted to resemblance and difference. Figurative language is constantly faced with the task of distributing attention amongst signs of likeness and unlikeness, both of which form an integral part of the powers of speech. The principle of unequal attention is already inscribed in how signs shown in the foreground relate to those thrown into the background; signs acting metaphorically thus entertain a close relationship with signs left unspoken because conveyed and "transported" via metaphor. Note here that the metaphorized is not a concept or mental image that exists outside the reticles of semiosis. It is rather just another sign (or signal) driven home by taxi in lieu of using its own means. The question is what motivates one sign to serve as the visible "vehicle" while the other (sign or signal) is reduced to a mere passenger (ibid.: 80–1)? Also, why the variations between one vehicle and another? Why use a metaphor instead of a simile, an allegory, a synecdoche, a metonymy, or a euphemism? Lastly, why should a figure of speech be preferred over "well-focused" denotations and proper names of literal communication?

Once again, answers to these questions can be found in derivations of our attentional theorem. As already suggested, a metaphor represents a precise mode of attentionality. It establishes a triangular connection between the metaphorizing s_1, the metaphorized s_2, and a determinate number of attributions of s_1, associations carved out from broader fields by way of implication and pertinence (the paschal lamb nomination implies innocence but not sheepish naiveté and simple-mindedness). The implied s_3, s_4, and "so on" come with s_1 (the metaphorizer) and are passed on to s_2 (the metaphorized). The resulting configuration is vertically ordered in the sense of distributing attentionality unevenly: none to extraneous associations (s^{-1}), more to s_1 compared to s_2, and more to s_2 compared to s_3, s_4 and s_n. Another feature of metaphorical assemblages is that they are done without the attentions of grammar, thereby maximizing closeness over distance. (Unlike the "French frog" metaphor, a proposition stating that "the French are like frogs" raises many questions!) The triangular, grammatically unmarked weavings of metaphor can serve a variety of purposes (humour, poetry, etc.), depending on the way the reticle is hooked up to broader Sr events.

It should be emphasized that language can deploy and combine a wide range of attentional modalities, well beyond the arrangements of metaphor. Sign processing can generate the narrow weavings of "straightforward" denominations, literal definitions, technical descriptions, logical inferences, and causal observations. These apparent exceptions to Ricœur's rule of plenitude point to the degree zero of rhetoric (ibid.: 140). They involve a no-style and no-digression tactic that differs from figurative language in one way only: the limited extent to which language casts its Sr net, to capture immediately relevant connections only, severing them from all other potential connections (especially those involving limbic affects). Zero-styled signs are not univocal references to objects and empirical reality that unite vehicles (driving words) and tenors (driven meaning) into undivided speech acts. Literal signs do not simply re-present objects and events, converting reality into words. More to the point, they are exercises in highly focused connectivity, synaptic configurations ruled by the most illiberal version of the principle of plenitude – one that will simply not allow signs to activate all compatible connotations. The attentions they generate are not free to deviate from discrete fields and levels of reticulation (be they pre-established or newly constructed; the potential for innovation is not the issue here).

Contracted reticles of the denominative, definitional, descriptive, technical, logical, or causal sort may be metaphors that are no longer alive, dead tropes that forego the playfulness of poetry, literature, and mythology (ibid.: 26f.). Denotations nonetheless have in common with metaphors the general principle that fashions their contribution to "living speech": the economy of attentionality. Everyday language is not merely a forgotten and used-up poem that no longer draws attention, as Heidegger (1975: 208) suggests. Rather, metaphors do what they do because of the way they hook up with words that are

attended non-metaphorically (Black 1962: 27). Contrary to
romantic, anarchic, and post-whatever views of freeplay in language (Ricœur 1977: 97), metaphors are not "superior" to
other usages of language. They are not even "superior" to the
ossified utterances of technical jargon and the hard sciences.
The Gadamerian notion (Gadamer 1994: 415, 431–3) that metaphoricity constitutes a motor force in living language (governed by the pursuit of order through the perception of
similarities) overlooks the powers of denotative attentionality.
While organized differently, with greater allowance for intra-
and inter-code foldings, *metaphor is nonetheless just another
cog in the wheel.*

Another problem with the concept of metaphor is that it
often glosses over important variations in the use of rhetoric.
Figurative tactics involve a broad range of attentional arrangements that cannot be reduced to metaphoricity. Consider other
tactics such as allegory, metonymy, synecdoche, euphemism,
irony, displacement, oxymoron, or plain ambiguity. *Allegories*
involve comparative parallels constructed between two recognizable fields. One consists of signs extensively attended by
speech (e.g., an animal farm story); the other consists of signs
implicitly attended via background associations (the history of
the Russian Revolution). Parallels of the allegorical kind trigger
evocative linkages within and between the two fields. They are
typically sustained in the form of a protracted narrative discourse (Orwell's book), usually with an *outward moral thread*
holding the story together (communism exchanges one tyranny
for another).

Metonymies and synecdoches are simpler tissues. A *metonymy* such as the Crown of England involves a proportional
analogy that projects an *implied difference* between ruler and
subject on to three distinctions: head versus body, superiority
versus inferiority, headdress versus full investiture. Verbal attention assigned to the royal headdress motif will suffice to activate
the entire reticle. The crown offers a centre of attention, the
sign upon which all implications converge. Contiguity is

another defining feature of metonymy. A metaphor differs in this regard; a man's "flame" carries a proportional analogy (the hot is to the cold what a lover is to a stranger) but no sign of contiguity. While rulers wear their crowns, lovers can act warmly without coming into contact with fire!

Synecdoches work differently. They usually draw attention to one component part within a larger object or event and treat it as a substitute for the whole. The sign thus singled out to convey predicative implications of the whole is elevated to the rank of *the one that matters or signifies the most*. English language illustrations of this attentional rank order device include a "sail" used for a sailing vessel, a "roof" for a house, "heads" for persons (as in per capita counts), "man" for all men and women, "length" for the long and the short, "height" for the high and the low, the "cross" for Christianity (cf. Barthes 1982: 211ff.), and so on. Alternatively, synecdoches can substitute the whole for one of its parts, thereby emphasizing a special connection between the two. For instance, the word "mortals" can be used to signify human beings, a substitution that carries two familiar implications: humans are central in the animate world, and the mortal condition afflicting all life forms is of greater concern to humans. Correlatively, ritual burial becomes a distinctive feature of human behaviour, turning *Homo sapiens* into *Homo funerarius*.

Euphemisms could be characterized as expressions that take some attention away from the inauspicious, distasteful, or offensive signs they imply, but not all of it. The tension built into euphemistic attentions is never fully extinguished. To speak of death as eternal rest, or life and death as the cycle of life, is to focus on the positive connections that can be projected on to what is still a mortal condition. Euphemisms thus "speak well" (Greek *eu-*, good, and *pheme*, voice, from *phanai*, to speak) of things we deem, somewhere "in the back of our minds," to be otherwise unpleasant.

Ambiguity differs from all previous arrangements in that it maintains a simple rule of divided attentionality. Words emitted or received ambiguously point to alternative pathways, an invitation to adopt one route only, and a dearth of information permitting choice.

Irony draws the attention to connections between signs spoken and contrary signs left unspoken. Things said explicitly are constructed in such a way as to emphasize contrary meanings that are fully noticed even though they are kept silent. *With irony, the unsaid is fully noticed and receives maximum attention.* Acts of speech that generate the incongruities of irony problematize and deny sign effects as soon as they are produced. For instance, feminist rumour has it that "when God [manifestly male, by convention] created man, "She" was obviously joking [divine femininity was implied from the start]." Irony produces the opposite of what is expected, reconfiguring sign connections in ways that contradict the apparent meaning of language or text.

Oxymorons play with polarities and contradictory connections as well, but with an effort to reconcile them at the explicit level. The spirit of mediation thus informs evocations of "a living death," "a thunderous silence," "a sweet sorrow," or "being cruel only to be kind."

Lastly, *displacement* and slips of language (e.g., feeling a sudden urge to talk about doughnuts in the presence of a police officer) also involve connections between explicit and implicit signs and signals. They act like hidden metaphors, so to speak. But the "hidden metaphor" expression is problematical in that it downplays surface differences that lie between the tropological and the non-tropological. As with rhetoric and figurative language, displacement plays on connections between the explicit and the implicit. But it also introduces another axis that *does not belong* to the science of tropology proper – i.e., signs operating on a level worthy of our fullest inattention.

I Like Ike

While bridging the gap between neuropsychology and semiotics, our Sr (sign-sign/sign-signal reticulation) theorem is a radical departure from stimulus-response and sign-representation theories alike. Signs are not linguistic responses to external stimuli. Nor are they representations of objects or concepts (that dwell outside the sign-signal process), words embodying ideas and responding to real circumstances and hermeneutic context. Our theorem emphasizes instead the weavings of signs-signals and the "attentional quanta" variably distributed among component parts of Sr activity. This approach has the advantage of broadening the notion of "sense" to include both practical syntheses of the senses and constructs of meaning, norm, and affect gathering in language. It also has the advantage of reconciling two basic rules of semiosis, those of conformation and fragmentation. One rule enables signs-signals to converge and form semiotic assemblages (the sound-image "ouch!" + automatic breathing signals). The other rule allows signs-signals to break up into separate connections that feed into new Sr activity (gasping continues after "ouch!" and ties in with memories of similar incidents). Both rules are essential to the nervous sign process.

The Sr approach throws light on variations in usages of language as well. Products of linguistic attentionality range from denotation and full grammatical expressions of likeness to tropes of all kinds. These variations in figurative arrangements should not be thrown into the glory hole of metaphor and polysemy. As already pointed out, a metaphor is one attentional tactic among others, with a reticular morphology of its own. The effect it offers is to draw the attention to one field or level of signification and assimilate it to another that remains implied but not fully expressed. Denotations work differently. They zero in on narrow sets of connections and pay no attention to playful sign potentials. A homonym reflects similar constraints; it requires that its reception be automatically confined to one field of acception only.

We have seen that grammatical markers betray attentional concerns as well. For instance, markers of likeness such as "resemble" or "to be like" can be elicited or elided with a view to expressing or silencing the reservations of comparative language. Attentional economics are also at work in rhetoric. Oratory forms differ in the way they distribute attention amongst fields and levels of semiosis. Each oratory form is characterized by the kind of connection it makes between an attentional centre and its immediate periphery, or the explicit and the implicit. Assemblages of indices and subindices of speech activity can be constructed through narrative parallels (allegory); projections of diacritic parts on to implied wholes (synecdoche); surface substitutions of the tasteful for the distasteful (euphemism); manifest tensions between divergent implications (ambiguity); deliberate contradictions between the said and the unsaid (irony); purposeful negations of contradictions (oxymoron); radiations from utterance to immediate associations (synonymous, polysemic); and so on.

Denotations, markers of grammar, metaphors, and figures of speech are so many products of attentional calculus. But they also point to another level of Sr activity that should be factored in and that has a direct bearing on the inter- and intrafoldings of sign connections: the normative and emotive overtones and undertones of attentionality. These axial considerations are particularly important to the infolding of signs and signals, wrapping them up in the foldings of judgment and affects of all kinds. They apply particularly well to the rule of metaphor. We have seen how the language of metaphor involves a cognitive weaving of explicit and implicit connections that specialize in attributions of likeness (with supplementary marks of *diakriti-kos* that can vary in amplitude). Although essentially correct, the argument neglects a critical dimension of metaphor. It leaves aside all connections between the licit and the illicit, hence the compatibility between things deemed alike and those judged or felt to be unlikable.

A good example of how likeness connects with liking can be found in Jakobson's discussion of a well-known slogan used in American politics of the 1950s: "I like Ike." The trisyllabic slogan brings together a subject, a verb, and an object. It also makes a paronomastic usage of the phonemic code. The utterance thus consists of:

three monosyllables and counts three diphtongs (ay), each of them symmetrically followed by one consonantal phoneme (..l..k..k). The setup of the three words shows a variation: no consonantal phonemes in the first word, two around the diphtong in the second, and one final consonant in the third. Both cola of the trisyllabic formula "I/like/Ike" rhyme with each other, and the second of the two rhyming words is fully included in the first one (echo rhyme), [layk] – [ayk], a paronomastic image of a feeling which totally envelops its object. Both cola alliterate with each other, and the first of the two alliterating words is included in the second: [ay] – [ayk], a paronomastic image of the loving subject enveloped by the beloved object (Jakobson 1985: 116).

Jakobson's analysis focuses on the phonemic and grammatical logic underlying the Eiseinhower slogan. The semantic aspects are left out. In a condensed manner, the trisyllabic formula combines signs of resemblance with ties of fondness; the former are achieved through phonemic technique, the latter through a single verb, "to like." The catchphrase also activates a distinction between two subjects, one identified by anonymous self-reference, the universal "I," and the other by nickname designation, namely, Ike. The distinction is politically slanted. While the "I" subject denotes a member of the electorate seeking representation through "aye" votes, the second seeks to be elected in order to "represent" all his supporting "ayes."

"I like Ike" brings together a show of euphony with signs of political liking. All is as if the pleasures of language were designed to reinforce electional preferences in the most natural way. "I," "like," and "Ike" have so much in common. The formula is a recipe for incorporating the prototypical subject named "I" ("ai") into a representative Ike ("aik") via the subject's supporting aye vote ("ai") and related signs of the like – signs of liking someone and being alike. Ike can be substituted for "I," "aye," and "like," each acting as a term that signifies other links in the chain.

The impact of the formula is to resolve all differences, acoustic and semantic. It does so by predicating "signs of likeness" on to connections and sentiments that bind the choosing and the chosen, the electorate and its elected representatives. The assemblage intermixes signs of elective fondness, representational resemblance, and referential "familiarity" (the self-familiar "I" and a man nicknamed Ike). The end product is a semantics of political harmony reinforced by notes of euphony and punning, a monosyllabic, alliterative, and paronomastic levelling of differences separating the American electorate from a candidate aspiring to the presidential office.

In its succinct way, the catchphrase highlights the substitutions of metaphor, a sign that points to something else it does not "properly" signify, invoking its presence indirectly, by way of resemblance. The same can be said of the playful ways of punning or paronomasia, from the Greek *para* (beside), and *onomazein* (to name), from *onoma* (name). It too names something with words that say something "beside" their proper meanings. Through the multiple meanings of paronomasia, the words "I like Ike" succeed in compounding effects of likeness, piling relations of similarity on sentiments of fondness and attraction.

Our deconstruction of an American slogan addresses the phonemic, syntactic, and semantic codification and mediation of similarities and differences performed on the cognitive level. It illustrates precise ways in which lines of divergence (LH) may be harmonized on a plane of convergence (RH). Mention has also been made of the political and emotive dimensions. Yet considerations of affect have been incorporated in a rather timid way. Our analysis mentioned the manner in which signs speak cognitively of feelings and judgment. This should not, however, be confused with the manner in which force and emotion are used in the handling of signs. How a slogan speaks about feeling and sentiment (irrespective of the tone accompanying the utterance) is a matter that belongs to the lateral domain, the area of cognitive activity (RH and LH). By contrast, the issue of how signs affect one another – how they struggle for primacy and attention – pertains to the axial dimension proper, hence the impact of norm and affect on sign attentionality.

This brings us to battles of sign attentions embedded in our American catchphrase. As already noted, both associations of liking and likeness are explicitly conjoined in the slogan. The same cannot be said of lines of divergence: they receive practically no attention. Why should this be so? After all, when situated in its proper context, the utterance presupposes struggles over issues that are eminently political. The electoral context against which this slogan is to be read implies social divisions and polemics as to how they should be tackled. One such division lies in the difference that separates a political office from the people it represents. Should voters support someone like them, or should they elect someone so admirable as to stand above all fellow citizens, distinguishing himself/herself before and especially after election? Another unspoken issue is the dividing line that separates the ayes from the nays, choices that are mutually exclusive and that entail expressions of both like and dislike. While foregrounding a two-sided effect of affinity combining resemblance and fondness, the slogan "I like Ike" brackets the two issues raised above, those of moral-political authority and electoral adversity.

Some signs are left in the background because a lot can go without saying, one might retort. Slogans cannot spell everything out lest they should err on the side of verbosity. But there is another reason that may account for the uneven attentions of "I like Ike," a reason that has to do with the actual effects that are being pursued. The point of the Eisenhower slogan is to draw attention to lines of convergence, away from divergences that may have divisive connotations. The end result consists in populist rhetoric – a display of signs of liking and likeness. The catchphrase hails a man known for his democratic simplicity and outgoing warmth. Ike resembles his fellow men in that he is of impoverished origins and is so familiar to everyone that people can refer to him by his nickname. The point, however, is that these effects can be attained on one condition only: that evocations of authority and political adversity be underplayed, *however conspicuous they may be in all other respects.*

Readers must bear in mind the context in which the slogan was heard. The most prestigious office was sought by a highly respected war hero upholding (moderately) conservative republican values, a candidate representing higher morals and presenting himself as a better choice over other candidates. Far from needing reinforcement, these evocations of "higher standing" were better left in brackets; electoral gains could be made by giving precedence to signs of populist rhetoric. A message of "representational affinity" is thus preferred over signs of moral-political verticality. The slogan brackets the authority vested in the "elected," passing over power differentials between voters and office-holders representing them. "I like Ike" is an invitation to vote for someone liked by everyone because he is like everyone, an imagery that counterbalances the entitlements of a hero and upholder of what people aspire to be, someone "like no one else."

At one level – the cognitive – the catchphrase speaks openly of affinities between voters and a given candidate. At another level, it *prefers to speak* about signs of affinity as distinct from indices of moral and political authority (obtained or confirmed through electoral means). This is the level of emotive attentionality proper, involving not words about choosy feelings but rather built-in feelings about choice words.

Through phonetic artistry, "aik" (Ike) is treated as a substitute for both "ay" (I) and "layk" (like). The illustration points to a central feature of all metaphorical utterances, which is to evoke or stand for other things they do not properly name. Our analysis, however, also points to several other phenomena, such as:

- the distance that lies between signs that do the standing (Ike) and those that are being stood for (I, like);
- the predicative connections that lie between the former and the latter (Ike is one of us);
- the normative injunctions embedded in the metaphor (be like Ike, vote for him);
- the affective underpinnings of language (wishing Ike's electoral victory); and, last but not least,
- the inattention applied to things that are better left unsaid (few are war heroes; few can aspire to the presidential office; few have the means to run for office; some don't like Ike because he thinks and is "like no one else"; etc.)

Metaphors serve to bring out relations of likeness in language. But similarities can be highlighted provided that subsidiary differences are retained in the background, with varying degrees of amplitude. The cognitive dynamics of metaphor thus operate on two planes simultaneously: the syncretic and the diacritic. To these lateral aspects of metaphor I have added two vertical considerations: the weavings of judgment, and the requirement that meanings deviating from particular morals be kept in check. The language of metaphor acts as an important cog in these wheelings of meanings, norms, and affects. It provides an overall direction to a given assemblage and offers a powerful venue for speaking to signs otherwise mute, unsettled, and disturbing. The task in question is achieved through the cunning ways of metaphor – signs chasing after objects of liking through relations of likeness.

In "I like Ike," language offers a tactic that imitates what it seeks but does not deliver. The slogan uses prosodic technique to attenuate the distance that lies between a presidential war-hero candidate and his electorate, emphasizing relations of likeness to promote expressions of electoral liking. Effects of liking are wishfully added on to effects of likeness. We have seen that metaphors can also go in the opposite direction. Instead of serving what might be called the *in-addition* effect, they can fulfil an *in-lieu* substitution function – as in shoes worn in missionary exile in lieu of the homeland property they resemble. The "I like Ike" strategy clearly opts against the simulational "in-lieu" strategy and offers instead signs of "adding-on" (a member of the circle of family and friends!), a syncretic tactic deployed with maximum amplitude. The candidate is to be elected not because of his exceptional qualities, attributes that place him in an ideal position to address issues of the day and to speak in lieu of those he represents, that is, those who are *pale reflections* of what Ike stands for. The in-lieu tactic reveals too much distance, an effect not in keeping with populist strategy. A republican campaign stands more to gain by conveying signs of closeness through wishful likeness and liking.

Interpretive and
Normative Judgment

What Aristotle says about metaphor can be extended to all acts of speech: they are constructed in such ways as to "set the scene before our eyes" (Ricœur 1977: 34; cf. 43, 58, 120). It is in the nature of signs to attract our attention whenever attention is needed. This can be done through rhetorical tactics of all sorts. But the staging of meaning can also be achieved through the foci of denotative and descriptive speech, depending on the effects and affects that are being pursued. As already suggested, high levels of attentional focusing – avoiding depolarizations going off in multiple directions – are vital to the works of denotation, description, definition, categorization, causal or technical explication, and so on. While fitting into broader webs of significations, sharply delimited sign connections are a *sine qua non* of human communication and survival. The polysemic fluidity and freeplay of semiosis is never so tyrannical as to preclude the effective delivery of simple lines such as "Watch out for the car behind you." Even poets will pay heed to the warning in its literal sense. Concepts of "literal meaning" or "intention" may be wrongly understood as a denial of the intricate weavings of signs and signals in language. But this is no reason to glorify semiotic freeplay to the point of ignoring the variable attentions that preside over figurative and non-figurative signs alike.

A direct implication of our Sr theorem lies in what is commonly called the "exercise of judgment" viewed as an act of sensible interpretation. Attentional focusing plays a particularly important role at this level – the interpretive deliberations and judgments of language. To give an example, readers will recall the hemlock imagery of the inaugural line of *Evangeline*: "This is the forest primeval, the murmuring pines and the hemlocks." My interpretive judgment was that the imagery can trigger memories of the poisonous Conium weed used to kill Socrates. At the same time, there is no point in denying that the primary connection conveyed by the surface script is between the hemlock and the Tsuga tree of the pine family. An interpretive judgment to the contrary is untenable. Likewise with the scriptural fig apron motif in Genesis. While the imagery can be hooked up to an edifying assemblage (as in stories of Zacchaeus climbing a fig tree for a sight of the Saviour), Genesis 3 chooses to highlight its sombre implications, tying it to the sufferings of human reproduction. The texts evoked here do not give equal attention to all possible ramifications of the hemlock and the fig. Some associative pathways are given less attention (through implication), or they may be completely blocked off from ground level activity (through inhibition). Communications will be closed off from attentional activity if and when they are incompatible with foreground and background indices of the script at hand.

Interpretive preference and normativity are corollaries of the *Sr* process. To use an older terminology, we might say that speech acts are replete with indices of literality and intentionality. But we have seen how problematical these terms can be. Concepts of the literal and the intentional convey "representational" views of signifiers linking up directly with what they signify, that is, perceptions of the real world (channelled through the senses) and related ideas or intentions of the mind. A theory centred on measurements of (in)attention can dispense with these metaphysics of words "representing" abstract ideas or concrete things. The attentional theorem nonetheless allows the sign process to impose limitations on how signs can be effectively assembled or disassembled. One such constraint lies in the necessity for speakers and listeners to exercise judgment vis-à-vis foreground and background connections deemed worthy of explicit or implicit consideration. Through applications of judgment, pertinent connections are duly attended and impertinent pathways duly ignored.

I should emphasize that the selective operations of judgment do not function in a purely cognitive or interpretive fashion. They also entail *effects of moral stratification*. By this I mean the moral grounds of *Sr* activity (involving prefrontal processing), a terrain permitting signs to elevate themselves above others, drawing attention to intimations of prescriptive goodness – signs of moral desirability and authority. The ethical ramifications of language have already been illustrated at length. Moral injunctions can be conveyed through simple utterances such as "I like Ike." Readers are also reminded of the "cycle of life" and measures of "luminosity" and "height" – expressions that convey the superiority of signs of life, light, and loftiness over imageries of death, darkness, and baseness. (For more elaborate illustrations of the interlacing of the cognitive and the ethical, see my analyses of naming practices, stories of beavers and frogs, or imageries of lambs and scorpions in Revelation and figs in Genesis.)

Effects of morality in language receive little consideration in semiotic theories of both structuralist and post-structuralist affiliations. More often than not, indices of normativity are treated as effects of deep/surface organization. They constitute conscious models that may or not account for rules built into the software codes of sign activity. Alternatively, norms are products of ideological rigidity (bourgeois ideology, patriarchy, eurocentrism, logocentric metaphysics, etc.). Morality in language is turned into an authorial/authoritative disposition forcing itself upon the terrain of semiotic freeplay, a form of bondage sustaining institutional fetters and related "chains" of signification. As Nietzsche once said, "the thing that *makes* an institution is despised, hated, rejected: men fear they are in danger of a new slavery the moment the word 'authority' is even mentioned" (cf. Heidegger 1968: 68). Models of authority now inspire as much fear as all expressions of "ideal models," exemplars and standards to be debunked through rebellious fits of postmodern "attitude."

One commonly cited target of postmodern "freedom fighters" is the Platonic conception of analogy involving ideal models based on scales of resemblance, for instance, the idea of courage and its multiple and uneven manifestations. The conception is criticized for its essentialistic assumptions. Platonic analogy links up signifiers with fixed essences to produce the hermeneutics of spiritual and theological anagogy. Aristotle's notion that metaphors act as mimetic devices designed to magnify and ennoble (as opposed to simply mimicking reality in language) is also to be faulted for its emphasis on the "upward motions" and "uplifting effects" of symbolic activity (cf. Ricœur 1977: 38–41). Critiques of this sort show such an aversion for everything that smacks of essentialism that they end up giving no serious consideration to the moral attentions of language. When signs of "judgment" are recognized, it is mostly in order to judge and denounce them ...

Can sign "values" be rescued from the structural and postmodern ruins of essentialism? I suggest they can. The rank order of sign values can and must be salvaged through mechanisms of selective and interpretive attentionality. These are the processes that determine those meanings that should be principally attended and those that should be either ranked second or debunked and removed from all sign offices. This rank ordering function should also be extended to moral attentions proper, or the exercise of judgment understood ethically. I am referring to the moral grounds of Sr activity, grounds that can be explored without metaphysical or metalinguistic pretension – without reverting to "essentialism." For instance, a canonical text such as Revelation shows how a lamb sent to the slaughter "reflects" on the Son of God and his supreme sacrifice. This is a lesson firmly situated in religious history. The implication of this "reflection" is not that the Lamb slain is merely a metaphor, a "pale reflection" of an archetypal model that hovers above its animal representation, as in anagogical thinking.

Moral judgment never transcends the text, freezing itself into eternal lessons that escape the actual history and idiosyncrasies of the "Great Code." Morality is an integral part of the weavings of sign regimes, not a set of principles informing semiosis from the outside.

The moral fabric of semiosis means that every locution comes with some allocution. In every logical code lies a code of honour; an ought dwells in every thought. All cognitive activities proceeding through hemispheric pathways entail some normative anchoring of sign projections mapped along vertical lines. Signs that simply try to point the finger at objects and thoughts are no exception to this rule; they too are prone to dictate things that ought or ought not to be.

Exploring the moral aspects of attentionality takes us beyond the logical taxonomies of tropes *à la* Fontanier, or the aesthetics of figurative language concerned with matters of style alone (ibid.: 10, 28–30, 52, 70–6). More generally, studies of normative attentionality are an invitation to move beyond all formulations of cognitivism, including the structural emphasis on codes of resemblance and difference. Correlations of binary oppositions *à la* Lévi-Strauss (1963: 213ff.) will not exhaust the subject matter of symbolling. A four-column representation of analogical correlations (a is to b what c is to d) projected on to the Oedipus myth will not do justice to the moral side of Greek tragedy. All products of language are deeply concerned with issues of right or wrong.

Moral affects operate not only through the devices of rhetoric but also the least formal and most ordinary forms of speech. Ethical considerations inform sign activity ranging from the conventional clichés, sayings, and proverbs of moral discourse to the most arbitrary assemblages of dream-like imagery. Prescriptive sign actions generate the embattlements of persuasion and sign acceptability, a weaponry of attentional techniques ranging from obscure metaphors and metonymies to clearer words of deliberation (political), judgment (judicial), and praise (epichtic). These battles are utterly pervasive: evaluative pathways are constantly used to determine not only things to be said that matter but also the manners in which they should be said. The substance of signs matters as much as their manners (more or less formal, more or less chaotic). Aristotle was right in this regard. What ought to be said is inextricably tied to saying it as one ought (cf. Ricœur 1977: 30–1).

Morality, Repression, and Transgression

The exercise of judgment has neurological foundations. It is a product of brain activity involving projections to prefrontal lobes. These connections enable the brain to control intense emotions such as fear or anger, keeping emotive and impulsive responses of the limbic system within bounds. Prefrontal lobes also intervene in step-by-step planning and goal-oriented actions, transforming limbic motivation into the orderly attentions of logic and reasoning. It is principally through exchanges between the prefrontal brain and the limbic system that the logical and the desirable interface, generating speech acts and behaviour governed by the pleasure principle (or the avoidance of pain) and rules of conventionality and ethics all at once.

Stress and injury can have deleterious effects on the interplay of prefrontal and limbic activities. Insults to the prefrontal pathways can have a negative impact on sociability, normal behavioural inhibitions, and goal-oriented action (producing excessive apathy or excitement). Under normal conditions, however, interchanges of logic and desire – the codifiable and the pleasurable – are a constant feature of human behaviour. Emotivity and rationality are deeply entangled. Axial connections between cortical and subcortical regions are so pervasive that familiar distinctions posited between rule-governed behaviour (prefrontal judgment and reasoning), repressive activity (limbic affects under control), and transgressive behaviour (limbic affects let loose) may be largely artificial. An alternative approach is to argue that judgment and morality in some way "contain" (convey *and* withhold) signs of transgression, which in turn "contain" expressions of the logic being transgressed. This is essentially the conclusion we have reached in our analysis of the ascetic and erotic underpinnings of foot and shoe fetishism. Paradoxically, norms and infractions are the warps and wefts of a single fabric.

Before we further unpack this thesis, more should be said about signs of repression and transgression. We have seen that the interplay of foreground and background attentionalities governs the cognitive and normative grounds of language. This is the terrain that expresses and impresses signs of what ought to be, signifying them as one ought. The ground-level pathways of semiosis, however, cannot be detached from another level at work in the order of desire: signs of the underground, "visceral" connections that are neither explicit (expressed) nor implicit (impressed) but rather illicit (repressed).

Signs of repression do not point to cryptic messages hiding below surface manifestations of language and behaviour. They point rather to pathways in need of energetic "capacitance" or closure. The pathways in question include all linkages that are near the attentional domain and threaten it to the point of requiring *systematic inattention* (obtained through reticular fragmentation, reduced frequency, or synaptic hyperpolarization, especially along the limbic-cortex highway). When inattention is deployed, impertinent connections closely tied or running parallel to attentional reticles are closed off. Relevant pathways alone are kept busy. Closing-off activity is needed to foreclose communications that can obstruct or contradict ongoing processes of moral and cognitive activity. Lines of the surface script can be facilitated on condition that digressive and offensive connections receive no impulse whatsoever.

The nervous sign process is a summation of countless acts of depolarization and hyperpolarization, a multitude of apertures and closures in semiosis. To use another terminology, tensions between ground and underground add up to the embattlements of an unresolved *différend*. As Lyotard (1988a: xi, 29) suggests, a *différend* consists of phrases of heterogeneous regimes in dispute, hence unavoidable encounters between the spoken and the unspeakable. The encounters are rife with shadow dialogues that never lead to final resolutions achieved through proper rules of litigation, clarification, and judgment. This means that lines written between signs are never decisive and straightforward. They never articulate positive connections alone, the kind conveyed through direct correspondences or equivalencies between signifier, signified, and referent.

A good dose of negativity is therefore built into lines of Sr activity. For one thing, lines of semiosis are signs of "no trespassing" – barriers against unwarranted rapprochements. In the words of Baudrillard (1981: 161 n.19), the negativity of the line "highlights what the sign denies, that upon which the sign establishes itself negatively, and of which it is only, in its positive institution, the symptom." Baudrillard goes on to say that denials built into language undermine all representational views of the sign process, pointing to non-places and non-values in the symbolic order. "To conceive the sign as censor, as a barrier of exclusion, is not to wish to retain for the repressed its position as signifiable, its position of latent value. Rather, it is to conceive it as that which, denied by the sign, in turn denies the sign's form, and can never have any place within it ... The symbolic is not inscribed anywhere. It is not what comes to be registered beneath the repression barrier (line), the Lacanian Sd. It is rather what tears all Srs and Sds to pieces, since it is what dismantles their pairing off (*appareillage*) and their simultaneous carving out (*découpe*)."

We know that signs of "no trespassing" are wasted if there is no compulsion to trespass. Repression is obliged to acknowledge the impulse to explore the non-places and the non-values of language. For signs to be silenced and closed off, there must be words and actions to that "obscure effect." Whatever language denies, "it will attempt to exorcise and integrate into its own operation: such is the status of the 'real,' of the referent, which are only the simulacrum of the symbolic, its form reduced and intercepted by the sign" (ibid.: 162).

This brings us back to the issue of displacement (e.g., saying "it doesn't matter" in Spanish and with an English accent, for no apparent reason), lines that deviate from sign-signal expressions and impressions and that serve neither logical nor tropological ends. The defining feature of signs "displaced" by others is that they involve a good dose of inattention, a mechanism that moves in the opposite direction of the attentional devices of logic, judgment, and rhetoric. Signs usurped by others are hyperpolarized by means of denegation. The act of usurping results in contradictory predications, the kind that pits foreground and background attentions against the underground ramifications of semiosis. The resulting silences, however, are never absolute. Shadows of loan translations and traces of distal pathways are inextricably woven into the fabric of immediate and proximal connections. Shadow dialogues produce transgressive lines at odds with the explicit and implicit contours of language; they generate off-track activity that can never be disentangled from signs of normativity. While hiding and resisting, immorality will always flaunt its powers of attraction and play an active role in pronouncements of the law. The end result is constant noise: paradoxically, morality is inclined to cheat and to show self-restraint.

Zoomorphic lines taken from the New Testament Apocalypse can serve to illustrate these points. Like any other corpus, John's "unveiling" of Revelation is full of apertures and closures. The explicit metaphors of the Lamb slain and the scorpion-tailed demons of Revelation point to implicit associations of Christ and the Antichrist. But they also serve to close off connections of "pagan" astrology, precluding celestial bodies of vernal Aries and autumnal Scorpio from acting as *bona fide* spirits worthy of John's attention. The metaphors of zoology are foregrounded and substituted for names of spirits uttered in the background (Christ, Satan). The exercise, however, carries a mission of far greater importance: forcing signs of astralism to go underground.

The expulsion of astralism is not absolute. Fragments of the sidereal cult are deviously recuperated, incorporated into the attentions of zoological metaphor. The co-optation is performed with discretion and circumspection, as it should be. Like metaphors, signs displacing other signs take on the appearances of "category mistakes." Unlike metaphors, however, the errors of displacement must be made with great caution, without being spotted. The impertinence obtained through slips of semiosis is no longer an artifice of surface rhetoric, a wilful category-mistake and violation of classificatory logic produced in order to shorten and strengthen a simile or an analogy. As a by-product of repression, displacement generates silent predications, offensive connections that cannot be the object of abridged or full-length comparisons. These mute predications can be expressed through gaps and crevices of language alone, turning silence into an another cog in the wheel of semiosis.

Revelation is a convoluted embattlement of biblical theology against "pagan" astralism. At the surface level, visions of the End pass judgment against men and creatures of sin. At a "deeper" level, judgment is passed against the cult of the visible spheres of heaven, asterisms marking the passing of time and the unfolding of human destiny and desire. The end result is a noisy mixture of the lawful and the unlawful; the overt and the covert cohabit in the foldings of Revelation. An injunction eminently worthy of Revelation – "thou shall not speak of astral gods" – is submerged in opaque symbolism, hiding itself in the secrecy of anagogy, never to be truly "unveiled." But to no avail. The transgressive comes out and assists in the teachings of virtue, via the sacrificial meekness of the vernal lamb, unavowed son of the equinoctial Ram. Logocentric morality (faith in the Verb alone) hides and cheats; immorality (astralism) rules and flaunts.

The rank ordering of signs is an alchemy of normative, repressive, and transgressive ingredients. These give flavour to all recipes of language; their impact is pervasive. To use a spatial imagery, the explicit and implicit follow routes that constantly invite illicit connections, linkages potentially prompted and triggered below ground – below signs of foreground and background attentionality. These invitations impact on the actual shape, contour, and details of overt Sr activity. They pressure surface patterns and signs into courting danger, maintaining contact with enemies of the attentional order, cohabiting and mingling with nomads travelling along unidentified paths.

Explications and implications of Sr activity never live alone, compelled as they are to seek the convolutions of language and desire. To use Lyotard's (1988a: 17) concept of *différend*, we might say that signs of the ruling order are like highly placed bureaucrats that send complicated commands to their subordinates, including edicts to partly disobey and interdictions to fully obey. Alternatively, there is Baudrillard's (1993a: 5) imagery of systems exploding beyond their limits, systems driven to override their own logic to the point of putting themselves at risk. Note that centrifugal propensities of the ruling process are not mere threats to system maintenance. In many ways risk-taking tactics verging on transgressive behaviour are essential to homeostasis. Decentredness is constitutive of the "logic of desire" – the will to power in language. In Revelation, the best way for Christ to silence Aries is by courting and appropriating the enemy's offensive lamb and ram imagery. For John, no risk means no gain, an outcome that spells ruin for the early Christian battle against the astrological "mathematics" of Rome, Egypt, and Babylon.

As Freud insists, moral authority thrives on the application of repressive energy, a mechanism that requires threatening forces to remain active and be worthy of repressive attentions. From a Derridean perspective, this means that one pole of an antagonism thrives on its opposition to another. This conflictual symbiosis thesis could easily turn against Derrida. That is, one could apply it to the debate pitting the metaphysics of presence/representation against Derridean grammatology, which emphasizes difference (as in the written *gram*, where signifiers never erase themselves or become transparent expressions of the psyche). Paradoxically, a graphocentric thesis developed on the fringe of western philosophy grants great force to its logocentric foe, a ruling enemy it must preserve against extinction if proclamations of triumph are to last beyond the initial victory. To treat a graphocentric axiology as having some elder's right over logocentric metaphysics has the effect of maintaining metaphysics by means of philosophical denial (Derrida 1981: 8–14, 21–6). An emphasis on difference derives meaning from its opposition to the metaphysics of presence.

Presumably, the same interdependence applies to any tension that develops between signs doing the displacement and signs actually displaced. Although essentially correct, the argument needs to be qualified. Briefly, the object of displacement does not consist of a peripheral force acting from the outside or the opposite side of the ruling order. As already explained, threats of transgression work on the Code from the inside. They intrude in the lawful pronouncements of language, generating affects that both weaken and strengthen dominant patterns of semiotic activity. Repressive affects thus require a force acting on both poles of an antagonism, a third term endowed with eccentricity and evading all binary conventions pitting one pole against another (Barthes 1982: 409), or centre against periphery. Eccentricity is in the centre.

The lawful "contains" (conveys/withholds) its own negativity, mixing together forces of repression and transgression to the point of turning the ruling order into a "crowned anarchy" at best (Deleuze and Guattari 1987: xiii; Deleuze 1994: 277). The word "process" can perhaps be used to capture these multiple and seemingly contradictory aspects of authority, repression, and transgression simultaneously at work in semiosis. We know that a lawful system or structure is always a "process," a way of doing and transforming things in an orderly fashion. But the term also entails procedures of conflict. Semiotic "processes" thus involve battles pertaining to contentious matters affecting the distribution of power and the proceeds of sign activity.

Nothing escapes these "processing" activities, not even subjects attempting to control them. Subjects may hold key positions, yet the activities they engage in also "process" them into subjects. A subject exists only by virtue of a will to power and unity within complex processes of sign and signal activity: "La notion d'unité n'apparaît jamais que lorsque se produit dans une multiplicité une prise de pouvoir par le signifiant, ou un procès correspondant de subjectivation" (Deleuze and Guattari 1987: 23–4). Although a centrally positioned player, the subject is a multiplicity of nervous fibres tied like puppet strings to other fibres forming a complex fabric. When set in motion, the fabric is called a plot. "Authors" or "agents" controlling motions of the strings from outside (ibid.: xvii) are neither true nor false to reality; they are merely characters woven into "metaphysical plots," subjects evolving in a mythical era called modernity.

As with all plots, repressive and transgressive effects do not last forever. Given the right conditions, both can be extinguished. Times and battles do change, whether through narrative motions or shifts in history. Rumour has it that modernity is no longer with us. Likewise, present-day astrology is no longer a serious threat to prophetic revelations of the End. Prophecy and divination are no longer at war. Signs of astrological transgression built into John's visions of the apocalypse have been extinguished; they no longer invite closing-off activity. While unthinkable at the time that Revelation was first written, reading Aries into the vernal Lamb slain imagery will hardly disturb present-day readers of the New Testament. When read into John's script, astrology no longer raises threats of imperial regimes echoing the evil forces of Babylon, Egypt, and Rome. Ancient astrological machines and their corresponding regimes have been disarmed and disassembled.

Can the same be said of "authors" and "agents" of the modern lifeworld? Let us see how Habermas and Husserl would answer the question.

Rationality and Lifeworld

Concepts of attentional economy and Sr processing permit a broad understanding of what the production of "sense" is all about. They are flexible enough to reconcile the two rules of conformation and fragmentation governing mutable assemblages of semiotic activity. The malleable distributions of sign-signal attentionality account for variations in usages of language ranging from denotation to metaphor and tropes of all kinds. Our Sr theorem also accounts for the normative and emotive overtones and undertones of attentionality, or the twofold interplay of the overt and the covert, morality and the logic of desire. Interchanges between normative and emotive (in)attentions point to effects of semiotic rank ordering, that is, sign attentions unevenly deployed in the exercise of judgment, be it of the interpretive, the instrumental, or the normative kind.

The rank ordering of semiosis permits the intermeshing of code and pleasure, rationality and emotivity – assemblages of explicit, implicit, and illicit Sr activity. Entanglements mapped along the axial plane are so pervasive that the "higher" expressions of judgment will "contain" the "lower" signs of transgression, and vice versa. Symptoms of these two-way "containment" effects can be found in tensions and shadow dialogues occurring between the said (foregrounding the Lamb slain motif), the unsaid (implications of asexuality in the background), and the unutterable (underground echoes of divine Aries). A corollary of this argument is that the lawful will exercise self-censorship. It will take measures to silence part of itself, the part that co-opts the unlawful into serving its own rule. Ambiguity and cheating are bound to ensue. Another corollary is that signs of transgression will partake in the teachings of morality. In the Judeo-Christian tradition, astromythology is contentious not because of its alleged licentiousness but rather because of the alternative teachings it offers. The Sabean offence (star worship) is to propose a worldly conception of motions of sacrifice, tying them to the visible rise and downfall of the great bodies in heaven.

My reading of astrological *mathematica* in Revelation (for a fuller version, see Chevalier 1997) underscores the importance of historical variations in discursive regimes and rationalities. But more should be said about the exercise of conscious rationality and instrumental judgment in semiotic activity. We have seen how acts of judgment involving interpretive and ethical assessments play an important role in semiosis. But this raises the question: What do we mean by judgment? Is it a separate faculty, a discreet function to be clearly demarcated from other aspects of *Sr* activity? Does judgment display constant features, such as rationality and discursive consciousness? Or can judgment be carried out through a multiplicity of forms, some of which may not be all that "conscious," requiring less declarative attention and clarity than others (as in anagogy)?

This brings us to concepts of communicational rationality and lifeworld developed in a Habermasian and Husserlian perspective. As we are about to see, phenomenology is useful in distinguishing conscious exercises in rational deliberation from social behaviour governed by automatic rules, values, and habits of social life. Using our terminology, we might say that more attention goes into producing an act of reason compared to a habitual lifeworld experience. This seems to confirm what phenomenology has to say about the same issues. The position adopted throughout this book differs nonetheless from the claims of phenomenology. While providing insights into these issues, Habermas and Husserl have a tendency to overemphasize the orderly properties of normativity and rationality. Their position is at odds with what this book has to say about fuzzy interchanges going on between judgment and desire, the licit and the illicit, the normative and the transgressive.

Consider the Habermasian dualism between deliberative judgment and lifeworld values. The Habermasian concept of rationality posits a necessary connection between judgment and consciousness. Briefly, Habermas holds that rationality may be either *cognitive-instrumental* or *communicational*. By cognitive-instrumental rationality is meant the employment of knowledge in goal-directed action, teleological behaviour generating practices of instrumental mastery. This form of rationality is prevalent in societies and cultures dominated by capital and represents a partial, if not distorted, understanding of rationality. Rationality should not be confined to this form. It goes beyond instrumentality to include communicational aspects as well (Habermas 1987: 66). By communicational rationality is meant the employment of propositional knowledge in assertions directed at overcoming views that are purely individual and subjective. This rational mode assures participants of the unity of the objective world and the intersubjectivity of their lifeworld. It is exercised with a view to reaching settlement by means other than the application of everyday routines or the use of force (or threats thereof).

Communicational rationality thus presupposes three things:

1   the constraint-free, consensus-generating force of the "better argument" (echoing the Piagetian principle of "social cooperation");

2   a storage of commonly shared assumptions and interpretations constituting a people's lifeworld – unproblematic situation definitions that provide background knowledge (implicit, prereflexive) for discursive practice and related deliberations of validity claims; and

3   the constant possibility that common understandings will not be reached; that consensus may have to be altered in order to be sustained; and that errors may be corrected and mistakes remedied (ibid.: 10, 13–14, 17).

It follows that rational validity claims are never automatically true by convention or by social currency. After all, value systems are not necessarily rational. Nor can we say of a system that is efficient in achieving system goals that it is consciously rational, except perhaps in a figurative sense. Principles embedded in the systemic functioning of society, *à la* Parsons, or in habits of social practice, *à la* Bourdieu, should not be confused with the exercise of reason and judgment. Communicational rationality hinges rather on discursive claims that may be defended or contested, hence means and methods by which claims can be refuted, modified, or grounded (backed up, evidenced, vindicated, redeemed). Good reasons and rational arguments constitute the court of appeal of communicational rationality.

Rational claims can take different forms. First, we have claims of theoretical discourse pertaining to issues of factual/objective truth. These involve knowledge claims, speech acts that generate descriptive judgments and serve to establish the existence of a state of affairs. Theoretical claims to truth also involve instrumental propositions pertaining to issues of means-end effectiveness, hence the efficacy of teleological action.

Second, reasoning can take the form of moral-practical discourse consisting of normative expressions and "ought" sentences. Moral statements are concerned with the rightness and acceptability of actions and norms of action and the extent to which they are morally judicious and practically reliable.

Third, the exercise of reason is part of explicative discourse, which aims at overcoming difficulties in comprehension, generating claims of "comprehensibility of well-formedness and rule-correctness of symbolic constructs." These claims speak to disputes about acts of speech, classification, judgment, or calculation.

Lastly, statements of aesthetic criticism and therapeutic critique should also be considered. Aesthetic criticism produces evaluative statements that concern the "good" and the "valuable" and that are sanctioned by social standards of value. Therapeutic criticism involves claims of subjective truthfulness or sincerity embedded in communicational activity, claims conducive to practices of self-criticism and discussions of authenticity and transparency (ibid.: 12, 23–30, 39, 41–2).

Habermas contends that rational conduct can be discussed without reference to rational enterprises and forms of expertise instituted in western history, be they juridical, medical, scientific, managerial, or aesthetic (Habermas 1987: 32, 36, 40). This is in keeping with our own "culturally disembedded" understanding of prefrontal judgment. Also, the Habermasian emphasis on the conscious operations of rational discourse can be reconciled with our uneven attentionality theorem, a calculus that permits high levels of normative focusing. Morality and judgment involve complex processes of litigation, deliberation, and adjudication sustained through high levels of connective attentionality.

Habermas tends, nonetheless, to overemphasize the *conscious* attentions, (claims, arguments, and validations) of communicative rationality. In doing so, he fails to detach his discourse on reason from western cultural forms. In keeping with Piaget's notion of "decentration" (from egocentric understandings of the world), Habermas argues that rational discourse may emerge provided that it is differentiated from subjective experience, objective facts, and norms of the social world (regulating interpersonal relations). Likewise, signs (Sgr) must be distinguished from their internal semantic content (Sd) and their external referents or objects in nature. Otherwise word and world are treated as one and the same. Words that are given magical powers leave no room for discursive criticism and revision of validity claims. For rational conduct to exist, worldviews closed and reified through the socialization of nature and the naturalization of society must be demythologized. This condition is not met in sociocentric expressions of mythical thought that follow the strict logic of similarities and contrasts, with little rational orientation to knowledge and action. Unlike mythical thought, rationality requires that persons and agents be distinguished from the objects they manipulate (ibid.: 46–52, 68–70).

Habermas and Piaget address a fundamental issue, which is the distance that humans perceive and construct between themselves and everything else that surrounds them and with which they strive to "communicate." This issue, however, should not be confused with another: the status of the subject/object distinction. Habermas and Piaget treat this distinction as a *sine qua non* of rationality. In doing so, they fail to recognize the

cultural specificity of the subject/object theorem, which is pivotal to the western "constitution." This is the theorem responsible for the notion that nature is devoid of intentionality; that
subjective ideas are essentially different from material objects;
that subjects are entitled to own objects (but not vice versa);
and so on. The subject/object distinction can also be found in
western forms of kinship constructed as syntheses of objective
biology (procreation) and subjective affinity (marriage), a language that most anthropologists rightly insist on keeping at
home. When tied to these premises, rational discourse will serve
to deliberate connections between subjects and objects. But
there is no reason to confine all forms of rational activity to
deliberative judgments framed by this subject/object theorem
and related practices and metaphysics of thought and thinghood (Heidegger 1962: 113). The notion that reason has to do
with "inferences or conclusions drawn from *facts* known or
assumed" is ethnocentric and cannot be demonstrated on rational grounds. The thing/thought theorem is an integral part of
a particular lifeworld – our own.

There is another problem with the Habermasian approach to
rationality: a tendency to overemphasize the reflexive and discursive aspects of judgment, reasoning, and morality. This
approach makes insufficient allowances for connections between
the rational, the pre-attentive, and the inattentive (hyperpolarized). We have seen how reason and logic may demand that full
arguments and injunctions be concealed or silenced if they are
to persuade and have an impact. This argument has already been
made at length and will not be repeated. Suffice it to say that
spelling out all aspects of an argument may be the worst way
to validate or challenge claims to truth, rightness, beauty, comprehensibility, or even sincerity. The insinuations and background operations of semiosis are no less important to morality
and rationality than all the pronouncements and deliberations
of declarative language.

Habermas sets the conscious exercise of judgment against the
background of a lifeworld consisting of implicit assumptions
pertaining to the world we live in. Against this view, we have
seen that not all elements of reasoning and claims to morality

(e.g., thou ought not to speak of astral gods ... oops!) require foreground attentionality. Some may be conveyed prereflexively and receive limited attention, as if they were part of a *lifeworld rationality and reasoning* better left unspoken.

The Husserlian distinction between pure theory and natural-lifeworld attentionality is faced with similar problems. As with Habermas, Husserl conceives the lifeworld as human will manifesting itself through collective habits. Each culture is a spiritual unity characterized by an intentional horizon or attitude. The latter represents a particular *telos*, "a habitually determined manner of vital willing, wherein the will's directions or interests, its aims and its cultural accomplishments, are preindicated and thus the overall orientation determined. In this enduring orientation taken as a norm, the individual life is lived" (Husserl 1965: 165). A culture's sense of directionality (*sens* in French) constitutes a people's lifeworld, an environing world governed by what is essentially a natural, mythical-religious, and primitive attitude. Husserl portrays this lifeworld as a normative orientation or style marked by closure and historical continuity, "a naively direct living immersed in the world ... that in a certain sense is constantly there consciously as a universal horizon." He goes on to say that "being genuinely alive is always *having one's attention turned to this or that*, turned to something as to an end or a means, as relevant or irrelevant, interesting or indifferent, private or public, to something that is in daily demand or to something that is startlingly new. All this belongs to the world horizon" (ibid.: 166, my emphasis).

Lifeworld experience is normative attentionality working on automatic pilot. It may be subject to variable levels of consciousness, piercing the world "with rays from the illuminating focus of attention with varying success" (Husserl 1962: 92). On the whole, however, reflexive and prereflexive truths emerging from this pre-scientific and pre-philosophical lifeworld are historically specific, perishable, finite, and inherently practical. The natural thesis that governs our wakeful living is filled with familiar facts and affairs, shared values, and day-to-day practicalities. It comprises a self-presence of the subject as lived-in cogito, "as someone who perceives, represents, thinks, feels, desires, and so forth" (ibid.: 94). Though called natural, this standpoint of what gives itself as existing out there is arrived at intersubjectively, becoming a "second nature," as it were. It is a taken-for-grantedness that we achieve as "we come to understandings with our neighbours, and set up in common an objective spatio-temporal fact-world *as the world about us that is there for us, and to which we ourselves none the less belong*" (ibid.: 95).

Husserl argues that consciousness exercised through the practice of theory differs from lifeworld thinking. The purely theoretical attitude developed as the overarching goal of the "united European will" and dating back to Greek philosophy is thus a radical departure from the pre-scientific attitude. With theory, the universal horizon of the lifeworld becomes a conscious *telos*. Unlike the lifeworld attitude, theory focuses its attention on the environing world in a manner that becomes resolutely "thematic," pursuing a culture of ideas and ideals that knows no limit. The end-goal of this culture of ideas is nothing less than a universal science of the world and the unity and well being of all beings. Scientific achievements are reproducible and imperishable. They create a never-ending movement towards an ideal image of the world and being, an eternal pole and infinite renewal of idealities (Husserl 1965: 157–60, 161–4, 166, 168).

This is not to say that theory and empirical science are the same. In his *Ideas: General Introduction to Pure Phenomenology*, Husserl (1962: 40–1) remarks that the theoretical attitude is profoundly unpractical, a deliberate *epoche* (Greek for pause) or bracketing of all commonly shared assumptions of the lifeworld and related practicalities of lived-in reality and the world-about-us. Transcendental phenomenology is a science not of real empirical facts but of universal essences and the non-real. Without denying or rejecting the world-about-us, pure phenomenology makes no use of our natural lifeworld experience and everything that we know about objects in our "immediate co-perceived surroundings." The lifeworld is nonetheless "*constantly there for me*, so long as I live naturally and look in its direction." This constitutes the natural standpoint, which can be backgrounded but never dissolved by the conscious or active adoption of another standpoint, such as the mathematical or the theoretical (Husserl 1962: 94).

Phenomenology chooses to bracket the natural thesis and all scientific standards and findings based upon natural evidence. It puts the natural attitude out of action but without doubting it. Husserl makes no concession to sophist, sceptic, or Cartesian doubts concerning the presence of what "still remains there like the bracketed in the bracket, like the disconnected outside the connexional system" (ibid.: 98). The Husserlian *epoche* is merely an abstention of the judgment *simpliciter*, *"a certain refraining from judgment which is compatible with the unshaken and unshakable because self-evidencing conviction of Truth"* (ibid.: 98–9). But why the abstention? The Husserlian answer is that an *epoche* is required to lay the foundations of a new scientific domain, a discipline designed to explore pure ego, pure experiences, and pure correlates of consciousness. The findings of this new science may corroborate truths placed in brackets and even take them as their point of departure. The latter can thus be accepted "in the modified consciousness of the judgment as it appears in disconnexion, and *not as it figures within the science as its proposition, a proposition which claims to be valid and whose validity I recognize and make use of*" (ibid.: 100–1).

Pure phenomenology underestimates the close ties that bind the discourse of theory to worldly attentionality. To begin with, second-order ideas about lifeworld attitudes do not have universal value in the sense of transcending the particular world-about-us, the one we live in and experience through events of cultural history. More often than not, the language of theory takes root in first-order ideas built into the lifeworld premises of a particular culture (e.g., assumptions regarding things deemed to be natural; the division between knowing/owning subjects and known/owned objects; the distinction between "ideas about things" and "things we think about," etc.). More importantly, the interplay between attentionality and inattentionality is complex. It cannot be reduced to a matter of progress, with some cultures showing off higher levels of consciousness and spearheading the evolution of humanity. Attentionality is a universal phenomenon not conducive to competition between cultures vying for superiority.

Another problem lies in the notion that shared assumptions of the lifeworld are finite and prone to a closure of the mind. Contrary to this view, we have seen that attentions to the world-about-us exhibit considerable playfulness, indeterminacy, and degrees of freedom that introduce some infinity into our daily horizons. Just as the openness of science and philosophy may be more apparent than real, so too the closure of historical lifeworlds to "other views" may be a western myth.

Lifeworld attitudes are not a simple matter of shared values and cultural consensus. Rather, they carry a cannibalistic taste for foreign elements and feed on routine deception and immorality as well (e.g., the Bible's virtuous "Lamb slain" imagery absorbing the heathenish signs of Aries). Lifeworld norms are full of silences and transgressions of their own, contradictory forces and affects that are poorly reflected in theories of "commonly shared attitudes and interpretations of the world."

Theories of consciousness that purport to generate self-awareness of the consciousness and the mind pose many problems. But should we not say the same thing regarding my own attempt to apply attentionality to itself (with a view to exploring the works of language and human behaviour)? Is neuro-semiotics not equally problematical? Perhaps. Theories that pay attention to attentionality offer nonetheless a novel advantage over phenomenological discussions of theoretical and normative consciousness: They never escape the lifeworld of body and perception. In the words of Merleau-Ponty (1974: 202, my emphasis), "Just as my body, as the system of all my holds on the world, founds the unity of the objects which I perceive, in the same way the body of the other – as the bearer of symbolic behaviors and of the behavior of true reality – tears itself away from being one of my phenomena, *offers me the task of a true communication*, and confers on my objects the new dimension of intersubjective being, or, in other words, of objectivity." In the end, the foundations of morality stem from the subject's attentions to other perceptions of his/her own perspective on things. In these mutual perceptions of distance lies the paradox of an alter ego sharing a common situation, an intersubjectivity that places "my perspectives and my incommunicable solitude in the visual field of another and of all the others" (ibid.: 210). More shall be said about the primacy of attentionality over consciousness, with an emphasis on the worldly, intersubjective, and self-reflexive features of attentionality.

Regimes of Desire

Students of sign activity should be wary of theories that simply denounce the strictures of verticality and stratification or the rigidity inherent in lifeworld semiosis. Playfulness in language and culture is highly compatible with the hierarchical operations of attentionality and fluctuations in the rank orderings of desire. In fact, semiotic fuzziness and malleability is a byproduct of the unequal distribution of attentionality, the will to power in language, the rank ordering of signs and signals, and the battles of morality and transgression in the exercise of judgment.

This position runs counter to some premises of postmodern philosophy. Deleuze and Guattari, among others, speak critically of the strictures of arborescent models of language and society, models based on concepts hierarchically distributed along vertical trunks and branches bifurcating at distinct levels of a binary apparatus. Models plagued with rigidities of this sort include linguistics *à la* Saussure or Chomsky and psychoanalytic pronouncements of the Oedipus dogma. Although suspicious of all great divides, Deleuze and Guattari see a fundamental opposition between arborescent forms of discourse (or social life) and chthonian systems of rhizomatic growth. Excurrent tree-like models are contrasted with multiplicities of reticles that have no principal roots and stratified divarications, underground growths that can move in new directions and territorialities with considerable freedom. Studies of rhizomatic fractals point the way to what these authors call schizoanalysis and the micropolitics and machinations of desire. The approach stresses the interplay of consistency and chaos, hence "chaosmotic" fields of consistency that are never reducible to planes of tree-like organization (Guattari 1995: 61, 64; Deleuze and Guattari 1987: 35, 64, 69).

Guattari (1995: 65–71) adds nonetheless that schizoanalysis must allow for variations in levels of organization or consistency. This caveat is not without consequence. It takes us back to a central thesis developed in this book: the malleable ways of "crowned anarchies" governing the economy of attention and the logic of desire. In a Deleuzian perspective the rule of "chaosmosis" means that weapons of chaos may be used in rhizomatic struggles against opinion and received wisdom (Deleuze and Guattari 1994: 204–6), a strategy I have already illustrated at some length (mostly with reference to frog imagery, pornography, and body-piercing). But our analyses (e.g., the "*no importa*" incident, animal motifs in Revelation, foot and shoe fetishes) have shown the opposite to be equally true: "chaosmosis" implies that subversive deviations can serve a dominant opinion. The fractal and the unlawful can be harnessed to regimes of desire.

The nervous sign process is neither simply arborescent nor absolutely chaotic. More realistically, our Sr theorem shows how sign activity combines cognitive relations with attentional investments, normative commands, repressive forces, and transgressive affects. The end result is a *regime of desire* that assembles all "machines and machinations" of Sr activity (Deleuze and Parnet 1987: 70, 104–6, 109). By definition, a regime of desire generates a problematic of its own. Persistent problems are distributed throughout a reticle or rhizome shaped by what Deleuze (1994: 280f.) calls the plane of consistency. This plane can turn pathological, as in dementia or schizophrenia, where obsessive iterations invest the entire psyche *in a single fragment*. Non-pathological regimes, however, require obstinate iterations as well. Differentiated series and levels resonate under the influence of a fragment or "dark precursor" traversing the reticle.

"Each series is therefore repeated in the other, at the same time as the precursor is displaced from one level to another and disguised in all the series" (ibid.: 291).

An example of this dark precursor lies in pagan asterisms (spheres of heaven treated as divinities) repeatedly denounced but never unveiled in Revelation. Shadow asterisms force John's attention to constantly wander off into foreign expressions of his own thoughts, alien grounds to be conquered and occupied as if his thoughts were still at home. In Revelation each series is thus "explicated and unfolded only in implicating the others ... Repetition in the eternal return appears under all these aspects as the peculiar power of difference, and the displacement and disguise of that which repeats only reproduce the divergence and the decentring of the different in a single movement of *diaphora* or transport. The eternal return affirms difference, it affirms dissemblance and disparateness, chance, multiplicity and becoming (Deleuze 1994: 300)." As with chaos and logos, alterity and familiarity are part of a single movement fed by the motions of becoming and eternal return.

Baudrillard's (1983, 1993b) political economy of simulacra provides another interesting discussion of chaos and fractality in sign activity. The author explores forms of "difference" constituted through simulacra, variable regimes or assemblages of desire spanning several periods of post-feudal history. As with Habermas, all regimes revolve around claims of truth, beauty, order, and authenticity. Baudrillard begins with feudal culture and classical art, a regime that problematized the battle between essence and appearance, truth and illusion, the real and the imaginary, the original and the counterfeit. The driving force behind feudalism lies in the genealogical apparatuses of roots, origins, and natural values (Chevalier 1997: 10–15). At stake here is the quest for origins and originality in art, language, science, and social position as well – signs of rank reflecting origins of status and class. Mercantile and industrial capitalism put an end to this order of simulacra, liberating signs from demands of rank and origins, whether social or natural. Capitalism instituted the law of exchange-value and made natural use-values responsive to market demand. Craftsmanship gave way to a mechanical reproduction of equivalent copies and the generalization of the commodity form (prefigured in the standardized architectural usage of stucco). Universal suffrage was to become the political expression of this new regime, a decision-making process plagued with problems of dullness and symbolic impoverishment (Baudrillard 1983: 83, 99, 104).

Signs of late modernity gave way in turn to cybernetic circuits, communicational digitality, genetic engineering, deep-structural modelling, and operational simulation. Programs of oppositional codes and binary principles now govern all living systems. Binarism prevails even in politics. Instead of deliberating social conflicts from multiple perspectives, public communications are regulated by electronic testing and digital polling. A public opinion massage and montage is manufactured through dual choice questions containing their own coded answers. Basic decisions are reduced to the either/or games of bipartite politics and reversible systems. Questions and oppositions are simulated, with answers and problem-resolution recipes already anticipated and controlled by the powers that be (ibid.: 105, 111, 125, 128, 131, 133). "From the smallest disjunctive unity ... [e.g., choosing between Coca-Cola and Pepsi] up to the great alternating systems that control the economy, politics, world co-existence, the matrix does not change: it is always the 0/1, the binary scansion that is affirmed as the metastable or homeostatic form of the current systems. Divine form of simulation ..." (ibid.: 135). The world embodies the mathematical spirit of Leibniz who "saw in the mystic elegance of the binary system that counts only the zero and the one, the very image of creation. The unity of the Supreme Being, operating by binary function in nothingness, would have sufficed to bring out of it all the beings" (ibid.: 103; see 152).

But there is yet another period in the evolution of western simulacra. Baudrillard (1993b: 140, 143) describes it as a fractal stage characterized by an epidemic of moments and fragments of semiosis. By this he means a hyperreal dissemination of value that negates the real (origins, equivalencies, codes) in

everyday life, injecting irony inside a reality constantly functioning against itself and heralding the closure of philosophy.
The stability and realism of previous regimes no longer hold.
Criteria are no longer available to evaluate the beautiful and
the ugly, the true and the false, good and evil, order and chaos.
Solidly grounded regimes give way to a haphazard proliferation
and dispersal of signs of value, flickering in and out of the
heavens of simulation. Signs turn into particles that follow
trajectories of their own, never to intersect. Fragments of previous orders continue but without their ideas, values, and aims.
Progress continues, yet the idea of progress disappears together
with all other grand narratives of modernity (democracy, science,
etc.) (Baudrillard 1993a: 5). Reality dissolves into simulation
and the hyperreal.

Paradoxically, Baudrillard's taxonomy of sign regimes is a
legacy of the third stage. Unwittingly, the author applies the
quest for codes and genetic forms to a bird's-eye view of western
sign history. Productions of simulacra are slotted into historical
forms driven by different codes, structuring laws, or genetic
principles. Baudrillard's last stage captures the notion that contradictions and multiplicities are inherent to the logic of desire
and the machinations of sign regimes. But the argument is over-
stated. The notion that fractality has now become the motor of
all sign activity is too simple and rigid a rule. Lines wandering
off in multiple directions never stray completely away from
regimes of similarities and differences, from norms and judgments that create values and reasoning, from regulations that
impose boundaries and blockages of all kinds, from subversive
activities that generate patterns and structures of their own,
from denials and distortions built into constitutional law, and
so on. All forms of semiosis are subject to these forces.

The outcome of these multiple forces can never be sheer
orgiastic confusion. Nor can it be the perfect indifference that
comes with the infinite sliding, formlessness, and ranklessness
of the nomadic sign. Semiosis evades the rule of groundless
negativity or Sartrian non-being, an indifferent black nothingness. Gaps of the synaptic or signaptic kind are not black holes.

They are sites of seething activity governed by openings and closures – by Heidegger's double movement of clearing and veiling, or Merleau-Ponty's process of folding and pleating. While activity in the gap engenders a world without stable organic identity, it is bound to create the orderly attentions and motions of nervous sign processing (cf. Deleuze 1994: 54–5, 57, 64, 276).

Mindfulness and Being-in-the-World

Judgment is a complex activity. Claims to knowledge, meaning, rightness, and reason are not reducible to simple typologies, be they couched in terms of lifeworld values, modes of rationality, forms of consciousness, or regimes of signs and desire. Nor are they amenable to the rule of chaos and fractality, rhizomes escaping all forces of rank ordering and the will to power in semiosis. Acts of judgment are never simple. For one thing they involve multiple levels of Sr connectivity organized along hierarchical lines. All claims of judgment are governed by a rank ordering principle involving the uneven attentions granted to signs of the normative and the transgressive, the explicit and the implicit, the licit and the illicit, the overt and the covert. Rank order is thus the cornerstone of any theory concerned with understanding the operations of judgment in language. It also happens to be a *sine qua non* of theories that purport to bridge the Kantian gap between morality and self-enjoyment, or suffering and pleasure (Lyotard 1993: 14–15; 1988b: 36, 40; Guattari 1995: 73–5f.). Hierarchy in signs may be relative in the sense of being necessarily constructed, yet the tendency to construct hierarchy is necessary and relatively unavoidable.

The principle of attentional stratification presents many advantages over theories of consciousness and mind. It comes nonetheless with a caveat: the fact that it draws its inspiration from neuropsychology. The anatomical origin of this concept gives it an air of intellectualism gone biological, reducing mind and world to brain and synaptic activity. Against this misreading of attentionality theory, I emphasize in the pages that follow the existentialist and phenomenological underpinnings of Sr theory, using concepts derived from Sartre, Merleau-Ponty, and Heidegger to bring out the inherent worldliness of semiotic activity.

One key feature of attentionality is that it does away with distinctions drawn between sign appearances (Sgr) and the essence of things and thoughts duly signified – objects and ideas supporting language from behind the scene. The argument echoes Sartre's position on matters of essence and existence. In *Being and Nothingness*, Sartre develops a phenomenological perspective gone existential and no longer "purely" transcendental. Distinctions between essence and existence are meaningless. "For the being of an existent is exactly what it *appears* ... The phenomenon can be studied and described as such, for it is *absolutely indicative of itself*" (Sartre 1956: 4). Assumptions of logocentric (Verb-based) inspiration no longer hold: for instance, the notion that infinity is attributed to an invisible Being and finitude to its visible manifestation. Infinity dwells rather in the visible world. Reality cannot be reduced to empirical perceptions alone or even their total sum. Each and every object is perceived in its totality. Yet this totality is grasped outside each and every appearance of it, without all possible appearances of the object ever being fully perceived and exhausted. Beings never reveal themselves completely. Appearances refer to other appearances and disclose a structure or potency. The structure in question does not hide behind the perceived object. It constitutes rather the principle of a limitless series of appearances and points of view that disclose the transphenomenal meaning of the object, not some noumenal essence understood by intuition and hiding behind its outward appearances. Infinity lies in the finite, and nowhere else (ibid.: 6, 8, 23–4).

How do we gain knowledge of this transphenomenal being or structure? Does knowledge of being hinge on interventions of the knowing and perceiving subject? The Sartrian answer to this question lies in an existentialist reformulation of the concept of consciousness, which is always *consciousness-of-something* (see also Merleau-Ponty 1962: 5). All acts of consciousness and practical judgments are directed towards and absorbed or "troubled" by objects and things that transcend them and that belong to the "outside" world. "Things" give themselves as already existing when revealing themselves to the consciousness that reveals them. Self-escaping transcendence is thus "the constitutive structure of consciousness ... born supported by a being which is not itself." The being of consciousness implies and thrives upon a being other than itself. Pure subjectivity that constitutes the beings of their own object and is severed from any external presence – from this "something" that consciousness is conscious of – destroys itself and disappears (Sartre 1956: 23).

Sartre's observations regarding consciousness can be extended to the concept of attentionality. As with consciousness, attentionality is not a unified and recognizable function that transcends its various applications, an intellect acting like a pure spirit hovering and lording over the body. Nor is it a faculty demarcated from other faculties in the same way that hearing differs from seeing (Merleau-Ponty 1962: 213). Attentionality underlies all acts of perception; yet it is never pure and can never free itself from being-in-the-world. It is always tainted by particular modes and the actual phenomena towards which our "attentions" are stretched (via writing, speaking, looking, running, etc.). Far from being reducible to acts of sense or products of the senses, these phenomena give themselves as already existing, acting "externally" in such ways that they can be tended or attended (and may be doing some attending themselves).

Sign attentions are inherent to the body. Merleau-Ponty is thus correct in locating the essence of sign-making in the body. The implication is that humans do not have senses that give them access to the world through sensations, to which is added ex-post-facto meaning by means of interpretation and thinking. Vision is not a physical mediation between mind and sensible matter, the thinking subject and the object of thought. Rather, vision places me in a certain field, a "horizon in action" where the activity of thinking is subordinated to the body. It is by reason of my position in that field that "I have access to and an opening upon a system of beings, visible beings, that these are at the disposal of my gaze in virtue of a kind of primordial contract and through a gift of nature, with no effort on my part; from which it follows that vision is prepersonal" (ibid.: 216; cf. 1974: 196, 200). In every act of seeing there is a horizon in action, a vision. Compared to this phenomenology of perception, the language of "concept" and "percept" is a rather cumbersome and superstitious account of the attentional dynamics of being-in-the-world.

Another point in common between attentionality theory and existential phenomenology lies in the notion of pre- and self-reflexivity. According to Sartre, all acts of consciousness require an immediate, non-cognitive consciousness of being conscious-of-something, a spontaneous relation or presence of the self and the subject to itself. "This spontaneous consciousness of my perception is *constitutive* of my perceptive consciousness. In other words, every positional consciousness of an object is at the same time a non-positional consciousness of itself" (Sartre 1956: 13). *The subject is necessarily conscious of it-self*, of the immanence of its own self-reflective distance, a phenomenon of consciousness that implies unity but not identity. Reducing the principle of unity to one of full identity and absolute selfsame cohesion or coincidence would cause the self to simply vanish (ibid.: 123–4). Sartre insists that the consciousness of the consciousness-of-something cannot precede or be given some sort of primacy over the consciousness of an object, as in Hegelian phenomenology (cf. Heidegger 1970: 112, 114). Nor should this consciousness be located on a higher level of abstraction and self-presence, at a distance twice removed from the world-about-us. Consciousness applied to itself is directly tainted by and cannot be severed from the particular phenomena "occupying one's mind." Sartre's view on this matter is radically different from Spinoza's concept of an *idea ideae* – subjects knowing that they know and knowing how to know *prior* to knowing anything else (Sartre 1956: 11–12).

The same observations apply to attentionality. We have seen that full reflective attentionality is not needed for the subject to attend the act of being attentive-to-something. A fleeting, background or prereflective apprehension of the act of attending-something may suffice. Actually, this "nonthetic" attentionality is a necessary condition of every act of attention and is immanent to it. As with Sartre's prereflective consciousness, self-attending attentionality is a precondition of the Cartesian cogito and cannot be derived from it. Words or numbers can be brought together in a meaningful way on condition that there be a unifying theme deploying itself as a self-attentive operation – the sustained act of speaking, writing, or counting. When attending an act, attentionality must be attentive to itself, if only half-attentively.

By contrast, nonthetic (prereflective) attentions can dispense with thetic (fully reflective) attentionality. Contrary to what Sartre says, a distinction can be made between experiencing something and "paying attention" to (being conscious of) the experience. Pain can exist "before" it receives our fullest attention (see *3-D Mind 3*); pain is detachable from the consciousness that we have of it. Yet feelings experienced without neocortical noticing still require neurological attendance. Attentionality is not an either-mental-or-physical phenomenon. Realism and idealism both mislead us into carving up what is an essentially indivisible event, a variable attentional chemistry that admits of no internal duality and mediation thereof (Sartre 1956: 13–14, 25–6, 121–3).

The subject of our attentional chemistry is always "in attendance" and is constituted as such. Paraphrasing Sartre, we might say that a being capable of self-attentive attentionality-to-something is a being that *is what it is not and is not what it is*. The for-itself of this "being-in-attendance" never coincides with itself in some full equivalence, a self-identity that contains no suspicion of some primitive crack, an emptiness or nothingness in its own being. Through this being-for-itself an impalpable fissure slips into being – impalpable in that nothing actually separates the subject from itself. It exists "in the form of an elsewhere in relation to itself," a presence to its own absence, a presence attending to the "world-about-me," a world located inside and outside all at once. As in reflective expressions of syntax, the sense of self marks a relationship and detachment between the attentive subject and its own self. The attentional being never acts as a being-in-itself, the kind that has an identity struck in one single blow. Self-attention is never so self-compressed as to know no temporality, no negativity, no otherness. The self can never be so full of itself as to be "*de trop* for eternity" (ibid.: 29, 120, 123, 126–7).

One immediate implication of this "attentional" rewriting of Sartrian phenomenology is that the fissure of the being-for-itself resides not in a metaphysical entity called consciousness. It lies rather in the anatomy of attentionality, a living organic site that calls for the "subject-matter" of neurosemiotics. Unlike Sartre's theory of "consciousness," the language of attentionality is firmly grounded in the phenomenal world and is an integral part of it.

Another important qualification of Sartrian existentialism concerns the ratio-based properties of attentionality. If anything, Sr activity is a plural function. It takes different forms and lends itself to complex admixtures of lights and shades applied to multiple phenomena. Attentional allocations are not reducible to a digital choice between prereflectivity (nonthetic) and full reflectivity (thetic). As already argued, attentionality works rather through variable sums and amplitudes of closures and disclosures. Phenomena can be attended to varying degrees provided that others are closed off and unattended. The end result is a composition of lights and shades, not a dual arrangement of perceptions and apperceptions.

Our light-and-shade argument applies not only to the phenomena we attend to but also to acts of self-directed attentionality. Much of what Sartre describes as nonthetic consciousness is the product of background attentionalities directed at "other things" that accompany and support foreground attentions. The act of paying visual attention to something thus requires two things: one, that we be aware of the fact that our eyes are at work; and two, that we pay less attention to this fact compared to the object we are looking at. We can always shift our attention, towards our eyes at work, taking some attention away from one thing (object seen) and giving it to something else (eyes seeing). But the more we become aware of our eyes focusing on an object seen at a distance, the less attention we pay to the object itself and the more attentive we are to the labour of perception (which is not the same as pure attention-ality, a ghostly phenomenon that cannot be attended and is without interest). This means that attention to attentionality is merely a sharing of Sr activity between two sides (internal/external) of the "world-about-me": the process of attending and the phenomena being attended, both equally physical.

The attentions we give the world-about-us never escape the body. This argument echoes Heidegger's understanding of the worldliness inherent to processes of the mind, a perspective that takes us further into issues of "mindful attentionality." As with Sartre's discussion of the consciousness-of-something, Heidegger insists that a sign is always a ready-to-hand *sign-for*. It is ready-to-hand in the sense of being readily available and obvious in its worldliness and everydayness. It is also a referential or indicative item of equipment deployed for-the-sake-of-some-thing, a serviceable for-something that signifies a "towardly" involvement. Equipment in a car consisting of flashing lights is for signalling a turn when involved in driving, for the sake of going somewhere – getting to the office to do some work, for example, or heading home and having dinner. As with car signals, all signs are part of broader equipment contexts (e.g., vehicles and traffic regulations) that exist prior to each and every sign involvement. These broader equipments give all the "assignments" and references of language their within-the-worldly character.

As soon as it is given a particular orientation within a deter-minate surrounding, sign action brings an equipment totality into the "circumspection of our concernful dealings." Through circumspection, the context of serviceability is properly sur-veyed and "takes over the work of letting something ready-to-hand become conspicuous," receiving attention within a mean-ingful surroundings or "environing milieu" (Heidegger 1962: 107, 109–12, 115–16; 1982: 164). When heading towards a chair in a room, we are never thematically conscious of one single thing only, say, the seat. Rather, the chair is part of functional surroundings, an intelligible thing-contexture already given beforehand and made up of walls, other chairs, stairs, windows, blackboard, neighbouring rooms and buildings (ibid.: 162–4). Likewise, a south wind that the farmer takes as a sign of rain is not merely a flow of air making itself present as it blows from a particular direction. It is not a simple occurrence that first makes itself available to apprehension and to which is then added a meaning or warning signal. "On the contrary, only by the circumspection with which one takes account of things in farming, is the south wind discovered in its Being" (Heidegger 1962: 112). Heidegger goes on to say that even when events come across as things that merely occur, they are apprehended as equipment that has not yet been understood, veiled as they are "from the purview of circumspection."

On condition that they receive our attention and are taken into consideration, signs and their context may be surveyed and made accessible and ready-to-hand for "our concernful circumspection." Unlike signs, however, a surrounding does not have to be made conspicuous in the sense of being discovered thematically. Actually, more often than not "the sign itself gets its conspicuousness from the inconspicuousness of the equipment totality" (ibid.: 112, see 114). A knot in a handkerchief may serve its purpose, which is to prompt a recollection, on condition that it does so at the right time and remains inconspicuous the rest of the time. The sign is part of a circumspective commerce with things that belong to a mnemonic environment that need not receive constant attention. The commerce is profitable provided that the knot acts as an equipment that has no intrinsic importance. Accordingly, it is forgotten as soon as whatever it marks is committed to an inconspicuous memory. It remains unnoticed until such time as the knot sign is duly attended and the signified brought back to memory. The knot points to the weavings of attention and inattention, signs of the foreground and the background tied to particular surroundings.

According to Heidegger, the work of "concernful circumspection" deployed in sign action offers clues as to how the "being" of the Dasein ("Being-There") and the world is inherently structured. Consider first the concept of "throwness." Being is always Being-with-others or Being-in-the-world; it is subject to constant motions of entanglement and falling. The Dasein is like a sign-for: it is inherently thrown into a world. "The Dasein must be *with* things" (Heidegger 1982: 161). Intraworldly existentiality marks the Self, generating a "they-self" that plants inauthenticity and alienation at the heart of the Dasein, a Being that flees in the face of itself, never to be simply its own Self (Heidegger 1962: 180, 227–9). In *The Basic Problems of Phenomenology*, Heidegger (1982: 137–8, 160) emphasizes this throwness of the self and the subject whose intentional relationship to the object is constitutive of the subject himself. The same cannot be said of the object qua object. The object is never concerned about how it relates to the subject. By contrast, "to relate itself is implicit in the concept of the subject. In its own self the subject is a being that relates-itself-to."

Intentionality is the care-structure or direction of the self-toward embedded in the Dasein relating-itself-to. A far cry from Kant's transcendental subjectivity, this relating-to "belongs to the existence of the Dasein. For the Dasein, with its existence, there is always a being and an interconnection with a being already somehow unveiled, without it being expressly made into an object. To exist then means, among other things, *to be as comporting with beings*." The everyday worldhood and concerns of the Dasein are such that "each of us is what he pursues and cares for ... As the Dasein gives itself over immediately and passionately to the world itself, its own self is reflected to it from things." The Dasein is never encountered or revealed through means other than genuinely and positively inauthentic (ibid.: 137, 139, 160–2). Transcendence is intraworldly, not otherworldly. It is an existential structure, a necessary condition for the subject to apprehend anything in the world.

To the extent that the subject is a being-in-the-world, then the world itself is a dwelling place that bears the mark of the intentional subject. The world is "a determination of being-in-the-world, a moment in the structure of the Dasein's mode of being. The world is something Dasein-ish" (ibid.: 166). As with signs-for, their equipment totality and their surroundings, the world consists of that which is ready-to-hand and immediately available to the circumspect subject. The world in question is already given and unveiled in advance, "thrown" at us well before we apprehend this or that. The world lies beforehand, prior to adding up the total sum of extant things. Extants that are "out there" are not for all that entities fully independent of the Dasein. They do not make up the "whole" of Nature and the universe viewed as formed matter, bearers of traits, or the unity of a manifold of sensations (ibid.: 163, 168–9; 1975: 30). Far from being extant, the world exists by virtue of a Dasein's intra-worldly being-with. It exists in a mode that is fundamentally Dasein-ish and that bears signs of the care-structure.

Given these considerations, the being of the world is best revealed in poetry and art. When painted by Van Gogh, "some particular entity, a pair of peasant shoes, comes in the work to stand in the light of its being" (Heidegger 1975: 36). In lieu of simply representing the shoes, the painting reproduces the thing's general essence, beckoning us to the true nature of the thing, placing it inside the intimacy of world and thing, in the separation and "dif-ference" of the between. This Heideggerian notion of dif-ference "does not mediate after the fact by connecting world and things through a middle added on to them. Being the middle, it first determines world and things in their presence, i.e., in their being toward one another, whose unity it carries out" (ibid.: 202).

In keeping with this Daseinish conception of art, the works of logic, language, and thinking are intraworldy activities that form an integral part of being. None of these works can be set against the beings or the world they address. They do not stand opposite to and come after the external objects of speech or thought they are said to "represent." In reality, the distinction between thinking and being, or *logos* (reason) and *physis* (nature, thinghood), is a stratospheric abstraction of western metaphysics. Things and thought are intimately connected. The nature of a thing always depends on a prior perspective, a line of sight that belongs to the Dasein and that is known in advance. Nothing can be said of clocks and celestial motions without a perspective on time, some prior knowledge of the reckoning and measurement of time. Etymology confirms the inherent worldliness of "thinking," a term evoking the act of having something "in view," "aiming at it" (Heidegger 1959: 116–18).

Not even logic can escape the worldliness of thought: it too is harnessed to *physis* by means of the gatherings of speech. Logic defined as the study of logos is not a schoolteacher's doctrine of thinking (Platonic or Aristotelian). It is rather a science of statement or discourse, as in *intelligere*, from *legein*, to speak (as in monologue or dialogue). This is to say that only when we speak (or engage in sign activity, we might say) do we think, not the other way around (Heidegger 1968: 16). The *legein* motif originates in turn from root words denoting the act of steady gathering, collecting, and reading (as in the German *lesen*), and hence relating one thing to another, as in "analogy." Thinking thus points to that which collects and endures in a state of togetherness, a primal gathering principle that stands at the heart of being. *Physis* and *logos* are one and the same (Heidegger 1959: 120, 122, 124–5, 127–31).

In the Heraclitean tradition, the gathering imagery implies the act of paying heed or attention to collectedness and the disclosure of true being and *logos*. The disclosure presupposes not only the togetherness of opposites moving and flowing into each other (e.g., life and death). It also presupposes the rule of rank, strength, and domination. By contrast, heedless hearing scatters and diffuses meaning, turning being into the rankless mass and turmoil of the many and the weak. It reduces being to the anarchy of dogs and monkeys. Inattention and inconsideration point in turn to a state of sleep or watchlessness, a form of mindlessness that gives credence to hearsay. Inattention is so mindless as to glorify mere appearances and opinions ruled by simple categorical and dualistic thinking (life is life and death is death) (ibid.: 131–3).

Although couched in a different language, Heidegger's discussion of worldly circumspection and intentionality reinforces much of what we have said about attentionality – that is, sign activity is a receptive way of attending to what is conveyed by the things we immerse ourselves into, and also an active response to their "call" or invitation. Calling something by a name, for instance "thinking," is our response to what directs, demands, beckons, or calls on us to reach out. The calling is for something to be reached by our own call (Heidegger 1968: 116–17). The response implies caring and tending. "What calls on us to think, demands for itself that it be tended, cared for, husbanded in its own essential nature, by thought" (Ibid.: 121). Not that what is called is the sheer product of an act of knowing performed through naming: while the calling summons something into a nearness, the product is still a "presence sheltered in absence." Things that are bidden through calling can never be fully wrested from their horizons of thereness and remoteness (Heidegger 1975: 198–9). The invitations of thinking come from what lies before the signifier and should be responded to by letting things lie before us just as they lie. "The bidding that calls things calls them here, invites them, and at the same time calls out to the things, commending them to the world out of which they appear" (ibid.: 200). The interplay of circumspective hereness and thereness underscores the invitations of mind and world. The mind perceives things "beforehand by taking to mind and heart. The heart is the wardship guarding what lies before us, though this wardship itself needs the guarding which is accomplished in the ... gathering." As with signs of "attending," thinking is "the minding that has something in mind and takes it to heart" (Heidegger 1968: 207).

The world exists by virtue of the surrendering and delivery of the attentions of brain, body, and language. At stake here is a gift of the self that converts thinking into thanking, from *thancian*, a word closely related to the old English *thencan*, to think (ibid.: 139). The Dasein does not exist prior to this gift of the self. To the extent that he behaves like a sign that reaches out to a world where it belongs, the signifier (both the sign and sign-maker) dwells in this calling. It is language that speaks and it is calling that offers us an abode (ibid.: 124, 216).

When seen in this light, signs are no longer instru-*mental* fingers that we use to point things out there in the world.

Rather "man *is* the pointer ... His essential nature lies in being such a pointer. Something which in itself, by its essential nature, is pointing, we call a sign. As he draws toward what withdraws, man is a sign" (ibid.: 9). The Dasein is a living assignment that does not take a fixed position along or outside the road of thinking. The signifier is not tied to a unique speech event that says something for one time only (*la parole*). Nor is it harnessed to a system of signs uniformly binding and available to everyone (*la langue*). More to the point, the signifier constantly remains "underway." To the extent that man is drawn into what constantly draws away, he points into the withdrawal, which means that the sign remains without interpretation. In the words of Hoelderlin, "We are a sign that is not read." In the final analysis, finding the answer to the question "What is called thinking?" requires that we always "keep asking, so as to remain underway" (ibid.: 10, 168–9, 191–2).

Given its "underway" nature, what is called thinking consists not only of the spoken word that receives our attention. It also includes the unsaid and a dialogue that "leads the speakers into the unspoken." This differs from a simple conversation that "consists in slithering along the edges of the subject matter, precisely without getting involved in the unspoken" (ibid.: 178).

Heidegger's care-structure captures many of the points already made regarding the nature of attentionality. It too converges on aspects of thinking that are so "embodied" as to be prepersonal, non-metaphysical, and non-representational. In its own way, the care-structure revolves around the "minding" process – tending to what must be attended with all the powers of our "larger intelligence," the body. While emerging from the *physis* of world and body, this "mindful disposition" lays the foundations of *semiosis*, converting all phenomena of the self and its surroundings into a "world-about-us." Some products of semiosis may give the appearances of things being apprehended "in-themselves," be they expressed through the language of the senses, common sense, or the natural sciences. Yet none of them can escape being constituted through the mindfulness of language, body, and brain.

Notes

1 This vertical hemispheric release hypothesis apparently tallies
with the fact that left temporal-lobe epileptics tend to exhibit
uncontrolled left-brain cognitive behaviour, with an excessive
preoccupation for ideas and moral or religious concerns.
By contrast, right temporal-lobe epileptics have less control
over their emotions, suggesting that unilateral lesions will cause
same-side emotional tendencies to be "released" (Liotti and
Tucker 1995: 403). Note also that subcortical atrophy and
low right hemisphere arousal patterns have been observed more
frequently amongst poststroke patients with major depressive
symptoms (see Bruder 1995: 665, 672).

2 Their findings regarding autonomic lateralization are somewhat
different. According to these authors, "left frontal and left basal
ganglia lesions lead to an increased frequency of major depression, while right orbitofrontal, basotemporal, basal ganglia, and
thalamic lesions are associated with mania. Anxiety disorders
without depression and psychotic disorders are also associated
with right hemisphere lesions, particularly posterior temporal
and parietal lesions" (Robinson and Downhill 1995: 708;
see Rosenzweig et al. 1999: 419).

3 Methods used to explore lateralized control of physiologic
functions include: (1) biochemical determinations of neurotransmitter contents of the brain (by means of high-pressure liquid
chromatography in bilateral dissections of postmortem brains
of normal subjects or experimentally lesioned animals, or by
means of PET imaging of neurotransmitter receptor binding in

the living human brain); (2) bioelectrical recording of event-related potentials (based on recorded electroencephalographic wave patterns, assessing either cardiovascular-related brain activity or electrocortical effects of neurotransmitter-related medication); (3) lateralized electrical stimulation of neural structures involved in the control of physiologic functions (e.g., peripheral autonomic pathways or regions in the lower brain stem or hypothalamus); (4) lateralized sensory stimulation using either short-term or prolonged visual exposition or noxious and pain-related stimuli; (e) lateralized elimination or inhibition of neural structures (either by unilateral brain lesions, unilateral ablation, or blocking of peripheral autonomic pathways, or by unilateral hemispheric inactivation using intracarotid amobarbital injection); and (5) measurement of physiologic effects of naturally occurring lateralized brain activation (by EEG power analysis or conjugate lateral eye movement) (Wittling 1995: 337–8).

4 Job 13.27, 18:5, 9, Gen. 3.15, 49.17.

5 Lev. 21.18, Deut. 15.21, Prov. 26.7.

6 Deut. 33.24, Job 29.6, Ps. 68.23.

7 Deut. 11.24, Josh. 1.3, 2 Chron. 33.8, Ps. 60.8, Eze. 29.11.

8 1 Chron. 28.2, Ps. 8.6, 31.8, 110.1, Matt. 5.35, Acts 2.35.

9 Isa. 5.26ff., Jer. 47.3, Ezek. 26.11, 32.13, Mic. 4.13.

10 Dan. 2.33, Isa. 22.23.

11 Num. 27.2, Josh. 19.51, Ruth 4.1, 2 Kings 4.15, Matt. 25.10, Luke 1.7, 12.36.

12 2 Sam. 6.5, 14–16, 1 Chron. 15.29, Ps. 30.11, 149.3, Eccles. 3.4, Jer. 31.4, 13, Luke 15.25ff.

13 Deut. 32.35, Job 13.27, 31.4–5, Prov. 19.2, Ps. 17.5, 94.18, Eccles. 5.1.

14 Prov. 1.15–16, 6.18, 6.27ff., Isa. 59.7, Rom. 3.15.

15 Ps. 2.9, Dan. 2.33f., Rev. 2.27.

16 Isa. 14.19, Eze. 6.11; Ps. 36.11; Isa. 49.23; Ps. 105.18.

17 Exod. 3.5, Josh. 5.15, Acts 7.33.

18 Job 38.8, 17, 41.1–34, Eccles. 12.4.

19 Deut. 22.21, 1 Kings 14.6, Acts 5.9.

20 Jer. 5.6, Matt. 24.43, Mark 13.33ff., Luke 12.37ff., Rev. 3.2–3, 16.15.

21 2 Sam. 4.12, Matt. 22.13, John 11.44.

22 Prov. 5.5–8, 9.14, Eze. 8.3, 16.

23 See also Eccles. 2.1ff., Luke 12.19–20.

24 Exod. 32.19, Judg. 11.34, 21.8–24, Job 21.11.

25 2 Sam. 2.18, 22.34, Isa. 35.6, Matt. 11.5, 15.30f., 21.14, Acts 3.7.

26 Exod. 3.5, 29.20, Lev. 14, Josh. 5.152, 2 Sam. 15.30.

27 1 Sam. 25.24, Luke 7.38, Acts 22.3.

28 Eph. 6.15; see Exod. 12.11, Rev. 16.15.

29 Deut. 8.4, 29.5, Neh. 9.21.

30 1 Sam. 2.9, Ps. 18.36, 40.2, 56.13, 66.6–9, 109.105, 119.59, Prov. 3.23, 4.26–7.

31 Lessons of virtuous devotion can also be conveyed by women's feet either refusing or accepting to dance. Readers are reminded of Vashti who refused to parade her beauty before a husband drunk as a lord and subjects making equally merry (Esther 1). But they should also bear in mind beautiful Abigail whose dance saved her drunken husband from David's vengeance (1 Sam. 25.36).

32 Exod. 33.8, 40.29, Lev. 8.3, 14.23, Num. 6, 12.5, 25.6, Mark 1.33.

33 Gen. 4.7, Exod. 12.6f., 22, 29.4, Lev. 1.5.

34 Acts 12.13, Hos. 2.15, Matt. 7.7, Rev. 3.20.

35 John 10.7ff.; see also Hos. 2.15, 2 Cor. 2.12, Rev. 3.8.

36 Isa. 3.16; see Deut. 28.56–66, Ps. 36.11.

Bibliography

Barthes, Roland. 1982. *A Barthes Reader.* New York: Hill and Wang.

Baudrillard, Jean. 1981. *For a Critique of the Political Economy of the Sign.* Transl. Charles Levin. St Louis, Mo.: Telos.

– 1983. *Simulations.* Transl. Paul Foss, Paul Patton, and Philip Beitchman. New York: Semiotext(e).

– 1993a. *The Transparency of Evil: Essays on Extreme Phenomena.* Transl. James Benedict. London: Verso.

– 1993b. *Baudrillard Live: Selected Interviews.* Ed. Mike Gane. London: Routledge.

Black, Max. 1962. *Models and Metaphors: Studies in Language and Philosophy.* Ithaca: Cornell University Press.

Boliek, Carol A., and John E. Obrzut. 1995. "Perceptual Laterality in Developmental Learning Disabilities." In *Brain Asymmetry,* ed. Richard J. Davidson and Kenneth Hugdahl, 637–58. London: Bradford; Cambridge, Mass.: MIT Press.

Bruder, Gerard E. 1995. "Cerebral Laterality and Psychopathology: Perceptual and Event-Related Potential Asymmetries in Affective and Schizophrenic Disorders." In *Brain Asymmetry,* ed. Richard J. Davidson and Kenneth Hugdahl, 661–91. London: Bradford; Cambridge, Mass.: MIT Press.

Burke, Stanley, and Roy Peterson. 1973. *Frog Fables and Beaver Tales.* Toronto: Lorimer.

– 1974. *The Day of the Glorious Revolution.* Toronto: Lorimer.

– 1981. *The Birch Bark Caper.* Vancouver/Toronto: Douglas and McIntyre.

Chevalier, Jacques M. 1990. *Semiotics, Romanticism and the Scriptures*. Berlin: Mouton de Gruyter.

– 1997. *A Postmodern Revelation: Signs of Astrology and the Apocalypse*. Toronto: Toronto University Press; Frankfurt: Vervuert.

– 2002a. *Half-Brain Fables and Figs in Paradise: The 3-D Mind 1*. Montreal and Kingston: McGill-Queen's University Press.

– 2002b. *Scorpions and the Anatomy of Time: The 3-D Mind 3*. Montreal and Kingston: McGill-Queen's University Press.

Chomsky, Noam. 1966. *Cartesian Linguistics: A Chapter in History of Rationalist Thought*. New York: Harper & Row.

Churchland, Patricia Smith. 1986. *Neurophilosophy: Toward a Unified Science of the Mind/Brain*. London: Bradford; Cambridge, Mass.: MIT Press.

Cohen, Ronald A., Yvonne A. Sparling-Cohen, and Brian F. O'Donnell. 1993. *The Neuropsychology of Attention*. New York and London: Plenum.

Daniel, Samuel. 1969 [1603]. *A Panegyrike with a Defence of Ryme*. Menston: Scholar.

Davidson, Richard J. 1995. "Cerebral Asymmetry, Emotion, and Affective Style." In *Brain Asymmetry*, ed. Richard J. Davidson and Kenneth Hugdahl, 361–87. London: Bradford; Cambridge, Mass.: MIT Press.

Deleuze, Gilles. 1994. *Difference and Repetition*. Transl. Paul Patton. New York: Columbia University Press.

Deleuze, Gilles, and Félix Guattari. 1987. *A Thousand Plateaus: Capitalism and Schizophrenia*. Transl. Brian Massumi. Minneapolis: University of Minnesota Press.

– 1994. *What Is Philosophy?* Transl. Hugh Tomlinson and Graham Burchell. New York: Columbia University Press.

Deleuze, Gilles, and Claire Parnet. 1987. *Dialogues*. Transl. Hugh Tomlinson and Barbara Habberjam. New York: Columbia University Press.

Derrida, Jacques. 1981. *Positions*. Transl. Alan Bass. Chicago: University of Chicago Press.

Driver, John, and Gordon C. Baylis. 1998. "Attention and Visual Object Segmentation." In *The Attentive Brain*, ed. Raja Parasuraman, 299–325. London: Bradford; Cambridge, Mass.: MIT Press.

Gadamer, Hans-Georg. 1994. *Truth and Method*. Transl. J. Weinsheimer and D.G. Marshall. New York: Continuum.

Goleman, Daniel. 1995. *Emotional Intelligence*. New York: Bantam.

Guattari, Félix. 1995. *Chaosmosis: An Ethico-Aesthetic Paradigm*. Transl. Paul Bains and Julian Pefanis. Bloomington and Indianapolis: Indiana University Press.

Guest, Edwin. 1968 [1882]. *A History of English Rhythms*. Ed. Walter W. Skeat. New York: Haskell House.

Habermas, Jürgen. 1983–87. *The Theory of Communicative Action*. Vol. 1, *Reason and the Rationalization of Society*. Transl. Thomas McCarthy. Boston: Beacon.

Hammond, Michael, Jane Howarth, and Russell Keat. 1991. *Understanding Phenomenology*. Oxford and Cambridge, Mass.: Blackwell.

Heidegger, Martin. 1959. *An Introduction to Metaphysics*. Transl. Ralph Manheim. New Haven and London: Yale University Press.

– 1962. *Being and Time*. Transl. John Macquarrie and Edward Robinson. San Francisco: Harper & Row.

– 1968. *What Is Called Thinking?* Transl. J. Glenn Gray. New York: Harper & Row.

– 1970. *Hegel's Concept of Experience*. Transl. Kenley Royce Dove. San Francisco: Harper & Row.

– 1975. *Poetry, Language, Thought*. Transl. Albert Hofstadter. New York: Harper & Row.

– 1982. *The Basic Problems of Phenomenology*. Trans. Albert Hofstadter. Bloomington and Indianapolis: Indiana University Press.

Heilman, Kenneth M. 1995. "Attentional Asymmetries." In *Brain Asymmetry*, ed. Richard J. Davidson and Kenneth Hugdahl, 217–34. London: Bradford; Cambridge, Mass.: MIT Press.

Hugdahl, Kenneth. 1995. "Classical Conditioning and Implicit Learning: The Right Hemisphere Hypothesis." In *Brain Asymmetry*, ed. Richard J. Davidson and Kenneth Hugdahl, 235–67. London: Bradford; Cambridge, Mass.: MIT Press.

Husserl, Edmund. 1962. *Ideas: General Introduction to Pure Phenomenology*. Transl. W.R. Boyce Gibson. New York: Collier.

– 1965. *Phenomenology and the Crisis of Philosophy*. Transl. Quentin Lauer. New York: Harper & Row.

Jakobson, Roman. 1985. *Selected Writings. Vol. 7, Contributions to Comparative Mythology. Studies in Linguistics and Philology, 1972–1982.* Ed. Stephen Rudy. Berlin: Mouton.

Johnson, Mark H. 1998. "Developing an Attentive Brain." In *The Attentive Brain*, ed. Raja Parasuraman, 427–43. London: Bradford; Cambridge, Mass.: MIT Press.

Kaite, Berkeley. 1995. *Pornography and Difference.* Bloomington and Indianapolis: Indiana University Press.

LaBerge, David. 1995. *Attentional Processing: The Brain's Art of Mindfulness.* Cambridge, Mass., and London: Harvard University Press.

Lane, Richard D., and J. Richard Jennings. 1995. "Hemispheric Asymmetry, Autonomic Asymmetry, and the Problem of Sudden Cardiac Death." In *Brain Asymmetry*, ed. Richard J. Davidson and Kenneth Hugdahl, 271–304. London: Bradford; Cambridge, Mass.: MIT Press.

Lévi-Strauss, Claude. 1963. *Structural Anthropology.* Transl. Claire Jacobson and Brooke G. Schoepf. New York: Basic Books.

Liederman, Jacqueline. 1995. "A Reinterpretation of the Split-Brain Syndrome: Implications for the Function of Corticocortical Fibers." In *Brain Asymmetry*, ed. Richard J. Davidson and Kenneth Hugdahl, 451–90. London: Bradford; Cambridge, Mass.: MIT Press.

Liotti, Mario, and Don M. Tucker. 1995. "Emotion in Asymmetric Corticolimbic Networks." In *Brain Asymmetry*, ed. Richard J. Davidson and Kenneth Hugdahl, 389–424. London: Bradford; Cambridge, Mass.: MIT Press.

Lyotard, Jean-François. 1988a. *The Differend: Phrases in Dispute.* Transl. Georges Van Den Abbeele. Minneapolis: University of Minnesota Press.

– 1988b. *Peregrinations: Law, Form, Event.* New York: Columbia University Press.

– 1993. *Toward the Postmodern.* Ed. Robert Harvey and Mark S. Roberts. New Jersey: Humanities Press.

Merleau-Ponty, Maurice. 1962. *Phenomenology of Perception.* Transl. Colin Smith. London: Routledge.

– 1964. *Signs.* Transl. Richard C. McCleary. Evanston, Ill.: Northwestern University Press.

– 1974. *Phenomenology, Language and Sociology: Selected Essays of Maurice Merleau-Ponty.* Ed. John O'Neill. London: Heinemann.

Nestor, Paul G., and Brian F. O'Donnell. 1998. "The Mind Adrift: Attentional Dysregulation in Schizophrenia." In *The Attentive Brain*, ed. Raja Parasuraman, 527–46. London: Bradford; Cambridge, Mass.: MIT Press.

Niebur, Ernst, and Christof Koch. 1998. "Computational Architectures for Attention." In *The Attentive Brain*, ed. Raja Parasuraman, 163–86. London: Bradford; Cambridge, Mass.: MIT Press.

Parasuraman, Raja. 1998. "The Attentive Brain: Issues and Prospects." In *The Attentive Brain*, ed. Raja Parasuraman, 3–15. London: Bradford; Cambridge, Mass.: MIT Press.

Parasuraman, Raja, and Pamela M. Greenwood. 1998. "Selective Attention in Aging and Dementia." In *The Attentive Brain*, ed. Raja Parasuraman, 461–87. London: Bradford; Cambridge, Mass.: MIT Press.

Parasuraman, Raja, Joel S. Warm, and Judi E. See. 1998. "Brain Systems of Vigilance." In *The Attentive Brain*, ed. Raja Parasuraman, 221–56. London: Bradford; Cambridge, Mass.: MIT Press.

Peters, Michael. 1995. "Handedness and Its Relation to Other Indices of Cerebral Lateralization." In *Brain Asymmetry*, ed. Richard J. Davidson and Kenneth Hugdahl, 183–214. London: Bradford; Cambridge, Mass.: MIT Press.

Posner, Michael I., and Gregory J. DiGirolamo. 1998. "Executive Attention: Conflict, Target Detection, and Cognitive Control." In *The Attentive Brain*, ed. Raja Parasuraman, 401–23. London: Bradford; Cambridge, Mass.: MIT Press.

Ricœur, Paul. 1977. *The Rule of Metaphor.* Transl. Robert Czerny. Toronto: University of Toronto Press.

Rioux, Marcel. 1974. *Les québécois.* France: Seuil.

Robbins, Trevor W. 1998. "Arousal and Attention: Psychopharmacological and Neuropsychological Studies in Experimental Animals." In *The Attentive Brain*, ed. Raja Parasuraman, 189–220. London: Bradford; Cambridge, Mass.: MIT Press.

Robinson, Robert G., and Jack E. Downhill. 1995. "Lateralization of Psychopathology in Response to Focal Brain Injury." In *Brain*

Asymmetry, ed. Richard J. Davidson and Kenneth Hugdahl, 693–711. London: Bradford; Cambridge, Mass.: MIT Press.

Rosenzweig, Mark R., Arnold L. Leiman, and S. Marc Breedlove. 1999. *Biological Psychology: An Introduction to Behavioral, Cognitive, and Clinical Neuroscience*. Sunderland, Mass.: Sinauer.

Sartre, Jean-Paul. 1956. *Being and Nothingness*. Transl. Hazel E. Barnes. New York: Washington Square.

Suliman, Mohammed. 1999. "The Nuba Mountains of Sudan: Resource Access, Violent Conflict, and Identity." In *Cultivating Peace*, ed. Daniel Buckles, 205–20. Ottawa: IDRC; Washington: World Bank Institute.

Swanson, James, Michael I. Posner, Dennis Cantwell, Sharon Wigal, Francis Crinella, Pauline Filipek, Jane Emerson, Don Tucker, and Orhan Nalcioglu. 1998. "Attention-Deficit/Hyperactivity Disorder: Symptom Domains, Cognitive Processes, and Neural Networks." In *The Attentive Brain*, ed. Raja Parasuraman, 445–60. London: Bradford; Cambridge, Mass.: MIT Press.

Swick, Diane, and Robert T. Knight. 1998. "Cortical Lesions and Attention." In *The Attentive Brain*, ed. Raja Parasuraman, 143–62. London: Bradford; Cambridge, Mass.: MIT Press.

Vale, V., and Andrea Juno, eds. 1989. *Modern Primitives: An Investigation of Contemporary Adornment and Ritual*. San Francisco: Re/Search Publishers.

Wittling, Werner. 1995. "Brain Asymmetry in the Control of Autonomic-Physiologic Activity." In *Brain Asymmetry*, ed. Richard J. Davidson and Kenneth Hugdahl, 305–57. London: Bradford; Cambridge, Mass.: MIT Press.

Wojcik, Daniel. 1995. *Punk and Neo-Tribal Body Art*. Jackson: University Press of Mississippi.

Zaidel, Eran. 1995. "Interhemispheric Transfer in the Split Brain: Long-Term Status Following Complete Cerebral Commissurotomy." In *Brain Asymmetry*, ed. Richard J. Davidson and Kenneth Hugdahl, 491–532. London: Bradford; Cambridge, Mass.: MIT Press.

Index